79 $\frac{50}{}$

Systems & Control: Foundations & Applications

Alexander Asachenkov
Guri Marchuk
Ronald Mohler
Serge Zuev

Disease Dynamics

Birkhäuser

Boston · Basel · Berlin

International Institute
for Applied Systems Analysis
A-2361 Laxenburg, Austria

Alexander Asachenkov
IIASA
A-2361 Laxenburg
Austria

Ronald Mohler
Dept. of Electrical
and Computer Engineering
Oregon State University
Corvallis, OR 97331

Guri Marchuk
Institute for Numerical Mathematics
Russian Academy of Sciences
Moscow, Russia

Serge Zuev
Institute for Numerical Mathematics
Russian Academy of Sciences
Moscow, Russia

Library of Congress Cataloging-in-Publication Data

Disease dynamics / Alexander Asachenkov . . . [et al.].
     p.   cm. -- (Systems & control : foundations & applications)
    Includes bibliographical references.
    ISBN 0-8176-3692-7. -- ISBN 3-7643-3692-7
    1. Immune response--Mathematical models. 2. Immune response-
- Computer simulation. 3. Tumors--Growth--Mathematical models.
4. Tumors--Growth--Computer simulation. I. Asachenkov, A. L.
II. Series : Systems & control.
QR182.2.M36D57   1993   93-40316
616.07'9--dc20   CIP

Printed on acid-free paper.   *Birkhäuser*

ISBN 0-8176-3692-7
ISBN 3-7643-3692-7

Typeset by the Authors in LaTeX.
Printed and bound by Quinn-Woodbine, Woodbine, NJ.
Printed in the U.S.A.

9 8 7 6 5 4 3 2

Contents

Preface

In this book we discuss the response of the immune system to antigen intrusion with the intention of establishing a basis for more effective treatment. This problem is of obvious significance to all mankind. Such a discussion is based on the assumption that clinical and experimental data have been accumulated, the analysis of which enables researchers to obtain the answers to questions of particular interest to them. However, in this analysis we need to use certain principles of mathematics as well as of biology. Until recently, the application of mathematics to biology, and particularly to immunology, was limited to elementary principles only. The methods of mathematical modeling which were successfully applied in other areas (e.g., the study of aircraft and rocket dynamics, atmospheric and oceanographic phenomena, etc.) were not used for the study of pathological processes. Perhaps this constraint is explained by the fact that there has not been a general conception or theory of disease upon which one could construct a corresponding mathematical model. Indeed, when we wish to construct a mathematical model of a complex dynamical process we need to distinguish and take into account only those factors which determine the reaction of the system to actions of concern within a particular study or experiment. Since the reaction of an organism to infection depends on a wide variety of factors, any attempts to solve this problem have had negligible success.

Clonal selection theory, which was a great stimulus for the development of immunology, provides the basis for considering different diseases immunologically and mathematically. According to this theory, which was confirmed by many experiments, each disease is considered to be the process of interaction between a pathogenic agent and the defense system of an organism via a unique chemical structure. In this way, the immune system utilizes its cells and molecules to kill and to neutralize the pathogenic agent. Viruses, bacteria, and tumor cells are such pathogenic agents. It is clear that, if viewed in this way, it is possible to

construct a mathematical model which describes the general characteristics of a very large class of diseases. The mathematical model allows
a variety of possibilities for the study of disease. They include, for example, the determination of the conditions of origin and the duration of
different forms of disease (i.e., acute, terminal, chronic), their stability,
the analysis of observed data, and the simulation of drug effectiveness
– all of which enable us to find the best way to treat the problem of
optimal control (for example, immunotherapy).

It is not surprising that all of these problems motivate the application of mathematics to medicine. A variety of models have been generated to describe processes occurring at the molecular and cellular levels.
Naturally, such models depend on the goals of the investigation. Monte
Carlo methods, stochastic differential equations, ordinary and partial
differential equations, as well as other mathematical theories, may be
used to model pathological processes.

At the beginning of our research in this area about 16 years ago we
found it most convenient and useful to describe a process using ordinary
differential equations. This is consistent with the biological principles
upon which immunology is based. It is of interest that the models derived from these principles are of a particular non-linear class. That
is, mathematically, the immune system and the diseases which it is
intended to control are simulated by coupled, bilinear systems. This
evolves in a natural manner from cellular and molecular conservation,
and from chemical mass–action principles. Consequently, bilinear systems (BLS) form a natural description of the dynamic interactions between pathogenic agents and the cells of the immune system.

This book is an integration of immunological principles, mathematical models, computer simulations, and methods of analysis. The main
notions and hypotheses of immunological theory necessary to understand
model derivations are given in Part I. The diversity of such models is
so large that it is impossible not only to study every model, but even
to enumerate them. Chapter 3 contains the most basic models, which
are composed of coupled, bilinear systems and which describe the immune response of the body to alien stimulation such as infectious disease. Chapters 2 and 4 are devoted to the development and study of
mathematical models of infectious disease.

Part II involves the derivation of methodologies which are useful for
model realization, parameter evaluation, and general system identification. This part of the book has a strong mathematical flavor relative

to other parts. The results obtained in Part II are applied to the solution of specific, real problems which are connected with the methods for processing clinical data. In 1975 some special functions, called gravity indices, were proposed for analyzing the course of such acute infectious diseases as viral hepatitis, pneumonia, etc. Experience in utilizing these functions has shown that they can be successfully used in solving such practical problems as an assessment of the patient's state of health, an analysis of the process dynamics, and a comparison of different treatment methods as far as their effectiveness is concerned.

Unfortunately, the use of such indices for the assessment of the state of oncological patients has not been successful. On the one hand, it is not possible to isolate an acute phase of disease when the parameters characterizing that process substantially deviate from their homeostatic values. On the other hand, it is difficult to isolate patients with a disease having a "smooth" course without aggravations and ending in a complete clinical recovery. All this suggests a need for the development of new mathematical methods of clinical data processing which take into account the specific features of the oncological process. These methods are discussed in Part III. The book ends with a set of conclusions, a look at new methodologies, and suggested directions for future research.

Alexander Asachenkov
Guri Marchuk
Ronald Mohler
Serge Zuev

The International Institute for Applied Systems Analysis

is an interdisciplinary, nongovernmental research institution founded in 1972 by leading scientific organizations in 12 countries. Situated near Vienna, in the center of Europe, IIASA has been for more than two decades producing valuable scientific research on economic, technological, and environmental issues.

IIASA was one of the first international institutes to systematically study global issues of environment, technology, and development. IIASA's Governing Council states that the Institute's goal is: *to conduct international and interdisciplinary scientific studies to provide timely and relevant information and options, addressing critical issues of global environmental, economic, and social change, for the benefit of the public, the scientific community, and national and international institutions.* Research is organized around three central themes:

- Global Environmental Change;
- Global Economic and Technological Change;
- Systems Methods for the Analysis of Global Issues.

The Institute now has national member organizations in the following countries:

Austria
The Austrian Academy of Sciences

Bulgaria
The National Committee for Applied Systems Analysis and Management

Canada
The Canadian Committee for IIASA

Czech and Slovak Republics
The Committees for IIASA of the Czech and Slovak Republics

Finland
The Finnish Committee for IIASA

Germany
The Association for the Advancement of IIASA

Hungary
The Hungarian Committee for Applied Systems Analysis, Hungarian Academy of Sciences

Italy
The National Research Council (CNR) and the National Commission for Nuclear and Alternative Energy Sources (ENEA)

Japan
The Japan Committee for IIASA

Netherlands
The Netherlands Organization for Scientific Research (NWO)

Poland
The Polish Academy of Sciences

Russia
The Russian Academy of Sciences

Sweden
The Swedish Council for Planning and Coordination of Research (FRN)

United States of America
The American Academy of Arts and Sciences

Acknowledgments

It is impossible to acknowledge the numerous contributions of all our colleagues, former professors, and students who have made this undertaking possible. Foremost, we are grateful to A.V. Balakrishnan, R. Baheti, G.I. Bell, C. Bruni, J. Dolezal, W. Duchting, Z. Farooqi, A. Gandolfi, E.V. Gruntenko, G. Hoffman, E.P. Hofer, T. Hraba, D.G. Knorre, G. Koch, A.N. Kurkin, A.B. Kurzhanski, K.S. Lee, Yr.M. Lopukhin, V.P. Lozovoj, S. Merrill, V.S. Mikhalevich, N.I. Nisevich, R.J. Poljak, A. Perelson, R.V. Petrov, R.I. Salganik, L.D. Sidorova, R. Strom, L.A. Trunova, N.N. Vasiliev, E.V. Zolotov, and I.I. Zubikova for their professional collaboration and advice.

Although virtually all our colleagues have contributed indirectly, there are several especially noteworthy contributors. These are L. Belykh, G. Bocharov, N.L. Grigorenko, C.S. Hsu, V. Janenko, D. Kalyaev, W. Kolodziej, I. Martianov, A. Pomanyukha, Ev. Smolianinov, B. Sobolev, and A. Yashin.

The Institute of Numerical Mathematics of the Russian Academy of Sciences, Oregon State University, and the International Institute for Applied Systems Analysis provided an excellent working environment and support. We are most grateful that our research on related topics has been sponsored by the US National Science Foundation and the Russian Academy of Sciences.

Introduction

This book summarizes the authors' experience in the field of theoretical immunology over the last 16 years. The main aim of the book is to demonstrate the practical application of mathematics to immunology – in which the process of an organism's defense against antigen intrusion is studied. Mathematical models of the immune response are developed on the basis of immunological principles. It was established that the reaction of the immune system to an antigen is independent of the antigen's specificity in relation to the pathological processes that were initiated by it in the organism.

A very significant contribution to immunology was made by F.M. Burnet, who proposed the clonal selection theory of antibody production. Subsequent research into the clonal immune response formed the basis for theoretical immunology. It should be noted that G.I. Bell was one of the first to describe clonal selection in mathematical terms. The numerical research of A.S. Perelson, C. De Lisi, C. Bruni, T. Hraba, G.W. Hoffman, R. Strom, and many others made a positive impact on the understanding of immune processes.

The field of mathematical modeling of the response formed an area of considerable research, with important advances by Mohler and his associates – in particular his research in bilinear system theory and its application to immunology. Such bilinear systems arise due to the nature of cell division, differentiation, and chemical binding.

The other authors have had many years of creative collaboration with immunologists, geneticists, biologists, and clinicians in Novosibirsk. This led to the organization of a seminar on mathematical models, which attracted the attention of a large group of prominent specialists in various branches of science. These factors made it possible to formulate a unified approach to the development of a model of immune response, its mathematical formulation, and the interpretation of the results. My first simple mathematical model of disease was written in Novosibirsk and developed by A.L. Asachenkov, L.N. Belykh, and S.M. Zuev. The research of the Novosibirsk scientific school was very influential in the evolution of the methodology of mathematical modeling.

It is not possible to mention all the research which has been done in this area and which has provided a solid base for the construction of more sophisticated models. Some of them the reader can find in this book and the accompanying references.

The disease dynamics theory was given new impetus by the author and R.V. Petrov when a more complicated model describing the reaction of the immune system to antigens on the basis of clonal selection and the co-recognition principle was proposed. This model was studied by A.A. Pomanyukha and G.A. Bocharov, who applied it to the simulation of viral hepatitis. It should be noted that this study was first introduced for clinical purposes, rather than mathematical ones, by N.I. Nisevich and I.I. Zubikova from the Second Moscow Medical Institute, in cooperation with I.B. Pogozhev and others from the Institute of Numerical Mathematics at the Russian Academy of Sciences. This is a brilliant example of a multidisciplinary project in which clinical and mathematical experiences were used to understand the process of disease in general and to provide a clue to effective methods for its treatment. Of course, these models are only a coarse approximation of reality, because such models describe an "average" patient. However, even in such a form, they enable us to bring various essential factors for the study of disease dynamics into the system. However, it is only the beginning.

Even though much progress has been made in theoretical immunology and the mathematical modeling of disease, many problems are still to be resolved. One of the main problems is the individual treatment necessary with respect to the immune status of the patient. This means that immune-status monitoring is an indispensable pre-requisite for the right choice of treatment. Thus, optimal control policies for disease therapy are of long-term interest: perhaps in the near future this will be the main problem. Clearly, the problem of how to describe the individual characteristics of the patient in mathematical terms is especially important when trying to use the model for decision making: for example, to choose the therapy or to forecast individual disease dynamics. There are many ways to solve this problem. One way was proposed by Pogozhev and developed by Zuev and Asachenkov. This approach is based on the concept of the similarity between different bodies and the micromovement intensity (see Chapter 10 for details) of particles (cells and molecules, etc.) in the liquid media (blood and lymph) in the organism. As a result, we are able to construct an integral characteristic of body vitality. This is a very interesting idea which allows us to study the correlation between different bodies. Using the available statistical information and so-called coefficient of similarity we can recalculate parameters of the model for a given person from those estimated for a group of patients. The preliminary results indicate that the proposed approach does not contradict clinical experience. Of course, it should be tested using other models and data. However, this approach looks very interesting and can be used for the solution of the optimal control problem, as well as other practical problems at individual and population levels.

Another important and extremely difficult problem is related to cancer research. It may be possible for the results which are presented by Asachenkov to be applied to clinical practice for the solution of such problems as the forecast of disease dynamics after surgery, on-line monitoring of clinically non-expressed tumor growth according to laboratory-measured indices, and the selection of an adequate scheme for treatment which takes into account the individual characteristics of the patient. To my knowledge, such results have not been available in the scientific literature until now. Of course, this is only the beginning of a long process in the field of cancer therapy.

Guri Marchuk

Part I

Models

Introduction

The present development of immunology is so rapid that even the basic concepts are changing right before our eyes, bringing into the arsenal still newer facts and hypotheses that may refine and even radically alter many parts of the theory. Nevertheless, the more basic principles of biochemistry upon which immunology is formulated can be considered established. This provides a solid base for the construction of models simulating the immune response and for new mathematical methods for clinical data processing.

Therefore, we briefly describe here certain immunological findings, facts, and theories with the aim of understanding the simulated process and the results of the modeling which follow. For more details see Burnet (1959), Nossal (1969), Roitt (1974), Petrov (1976,1983), Gruntenko (1987), and the *Annual Reports of the Basel Institute for Immunology (1987–1991)*.

The immune response is related to the universal character of an organism's defense against bacterial and viral attacks, as well as against poisoning by products of viral–bacterial activity, or intoxication by foreign agents of a biological nature. Although we shall not be able to discuss most of the recent advances in immunology here, throughout Chapter 1 we shall discuss the basic properties and concepts of the immune system with the aim of showing that both theoretical and practical problems of considerable importance in immunology cannot be solved by purely experimental techniques. They require the development of mathematical models. The most important discovery of the last few years was the establishment of the fact that the reaction of an organism's immune system to antigens is independent of its specificity in relation to the pathological processes that are initiated by antigens in the body. Therefore, knowledge of the immune response mechanism provides a clue to understanding the processes of diseases in general, and methods for their effective treatment.

In Chapter 2 we study the simple mathematical model of a disease which was proposed by Guri Marchuk in 1975. This model is only a crude approximation and generally requires further refinement. However, even in this form, it enables us to bring various factors essential to the dynamics of an infectious disease into a system, and it is useful for studying the general picture of the course of a disease and clarifying some observed results.

A more detailed study of the immune response suggests the construction of more complicated models. Therefore, in Chapter 3 the humoral, clonal-selection, and cell-mediated models, and their bilinear synthesis and analysis, are studied by Ronald Mohler. The model describing the reaction of the immune system to infectious-agent invasion, on the basis of clonal-selection theory and the co-recognition principle, is considered in Chapter 4. The equations describing the model were proposed by Guri Marchuk and Ram Petrov. Organ-distributed immune dynamics are studied in Chapter 5, following the work of Ronald Mohler and his associates.

Chapter 1

A Basic Overview of Immunology

The immune system is the set of lymphoid organs and cells that contains the thymus, spleen, liver, lymphatic nodes, Peyers's patches, lymphocytes of bone-marrow derivation, and the peripheral blood. All these represent a connected "diffuse" organ with a mass of about 1.5–2.0 kg. The total number of lymphoid cells is approximately equal to 10^{12}.

The main aim of the immune system is to defend an organism from agents with properties of genetically alien information (such as bacteria, viruses, proteins, tissue, and the organism's transformed cells such as tumor cells). The immune system generates cells and molecules to bind and to destroy the alien. These defenders are circulated throughout the body organs by the bloodstream and throughout virtually all tissue, and they are appropriately processed in their migration and recirculated back through the lymphatic vessels.

The generation of these defender cells and molecules is termed the immune response. Any alien substance which is able to induce such a response is called an antigen (Ag). An antigen can occur in a free state, as well as on the surface of cells. A single antigen may carry several distinct molecular groups (epitopes), each of which, in cooperation with others, may stimulate an immunological response.

There are two main types of immune response (in addition to the non-specific response which, in general, we do not consider here):

- Humoral immune response – immunological defenses that are determined by antibodies generated from B-cell dynamics.
- Cell-mediated immune response – immunological defense against foreign agents which is mediated by various types of lymphocytes. This manifests itself, for example, as the rejection of an organ transplanted from one animal to another. Such a response is supplied by the T system of immunity. Macrophages and natural killer cells are other important components of cell-mediated immunity.

Two main aspects of the immune system's response are its specificity and memory. Specificity means that antibodies produced in response to a certain chemically specific antigen are able to bind this antigen strongly, but are not able to bind other antigens so strongly. The second aspect is memory. Usually, the specific recognition of a given invader leads to "memory" in the immune system. A second invasion by the same infectious agent provokes the system to respond more effectively and results in an accelerated elimination of the infectious agent. This is the basis for vaccination against many infectious diseases.

Looking at the situation from the other side, the immune system's response can be manifested in allergic reactions in which the same specific recognition make us feel ill. Defects in the recognition and cooperation process between the cells and molecules of the immune system can lead to immunodeficiencies (Lopukhin and Petrov, 1974, 1975). Perhaps the most terrible example is AIDS.

1.1 Antigens and Antibodies

An antigen usually enters the epithelial layers of the skin, lung, or gut. A virus may multiply in these cells, often changing their properties, and may even display part of its structure on the surface of the epithelial cells, which can be recognized as foreign by cells of the immune system. Bacteria, and many non-multiplying antigens, may reach the bloodstream directly and be carried to the lymph nodes and spleen where they are recognized as being foreign by the molecules and cells of the immune system.

A particular class of white blood cells called lymphocytes are responsible for immunity. These lymphatic system cells circulate in the body and are present in high concentrations in certain lymphoid organs such as the spleen and lymph nodes (Sapin, 1978).

Proteins produced by lymphoid cells in response to antigen penetration in the body are known as antibodies (immunoglobulin). In turn, such antibodies can react with the antigen that induced their production. As a result of such interaction, immune complexes (antigen–antibody) are formed and removed from the body. There are at least five major classes of immunoglobulins: IgM, IgG, IgA, IgD, and IgE. In normal human adults approximately 75% of serum immunoglobulins are IgG. The IgM class, consisting of about 10% of serum immunoglobulins, is prominent in early responses to most antigens.

Antibodies, or immunoglobulins, are so crucial to the immune process that it is important to discuss their properties in more detail. They are molecules that, through binding sites, may bind to certain chemically defined small portions of the antigen called antigenic determinants or epitopes.

Typical antibodies, IgG, are Y-shaped protein molecules with a molecular weight of 150,000. The molecule has one constant region, C, and two variable V regions. The organism is capable of producing many antibodies of, say, 10^6–10^7 differences in structure according to the V region. The V region is a three-dimensional formation that to some extent is complementary to the antigen that induced the production of a given antibody. The antibody uses the V regions for antigen binding which simplifies its removal from the body. While antibodies possess high specificity in relation to a given antigen, they do not form an absolutely homogeneous population. They differ from each other in their affinity for a given antigenic determinant. Related to this, there exists the notion of antibody avidity, i.e., its ability to generate a strong bond with the antigen. Such a bond is reversible. The bonding strength is called affinity and can be calculated using measurements of association constants. The higher this constant is, the stronger the bond.

1.1.1 Example: antigen–antibody interaction

One of the first models for antigen–antibody interaction was developed by Bell (1973). Suppose that each antigen (Ag) has only one receptor for binding to an antibody (Ab), and each molecule of antibody can bind only one antigen. In fact, an antigen has several sites for binding with antibodies, and an antibody has two (IgG) or more sites for binding with an antigen. The model describing the interaction between antigen and antibody in terms of using a predator–prey type of relationship in the conservation equations becomes

$$\frac{d}{dt}\mathrm{Ag} = \lambda_1\mathrm{Ag} - \alpha_1(\mathrm{Ag})_b$$
$$\frac{d}{dt}\mathrm{Ab} = -\lambda_2\mathrm{Ab} + \alpha_2(\mathrm{Ab})_b\left(1 - \frac{\mathrm{Ab}}{\theta}\right) , \qquad (1.1)$$

where $(\mathrm{Ag})_b$ and $(\mathrm{Ab})_b$ are the concentrations of binding antigen and antibody, θ is the limiting concentration of antibodies for a given species, and λ_i and α_i ($i = 1, 2$) are appropriate constants.

This type of model is called a point model or a model of perfect intermixing. The hypothesis of perfect intermixing supposes that at each instant in time, for all points of the system, the concentration of a given substance is equal. This is a good assumption if the spatial-averaging of reactant concentration throughout the mix is much faster than the chemical reactions. The correctness of the hypothesis for the immune response depends on a large number of interacting units.

1.2 Lymphocytes

Antibodies are produced by a type of white blood cell – the B lymphocyte. These cells are almost morphologically indistinguishable from another class of lymphocyte, the T lymphocyte. T cells can also interact with B cells and play either a helper or suppressor role in regulating the immune response. Formally, we can divide the population of T cells into the following classes: T-helpers, T-suppressors, and T-killers (Petrov and Dozmorov, 1985).

Lymphocytes derived from bone marrow stem cells play a central role in the immune system. The stem cells are the common precursors of all blood cells: erythrocytes, granulocytes, megakaryocytes and platelets, monocytes and macrophages, and T and B lymphocytes. It is important to realize that lymphocyte development does take place throughout life, not just once during embryonic development, because 0.1–1.0% of all lymphocytes in the system die daily and must be replaced.

Lymphocyte differentiation occurs at special sites in the body. Stem cells produced in bone marrow circulate in the blood and penetrate the thymus and other lymphoid organs (Chertkov and Fridenshtein, 1977). In these organs, stem cells are transformed into lymphocytes. This process is accompanied by the multiplication and accumulation of lymphocytes. In the thymus this transformation leads to T lymphocytes. Although T cells may have a common precursor, their repertoires and functions appear to be different, in that they are killers at one site, and may be helpers or suppressors at other sites in the immune system. If this transformation takes place in the bursa in birds, or possibly its analog in mammals, then B lymphocytes are formed.

1.2.1 Receptors

Let us now look at the receptors. A receptor is a molecule inside or on the surface of a cell which recognizes a specific hormone, growth factor, or other biologically active molecule. The receptor also mediates the transfer of signals within the cell. The three types of receptors on different types of cells have to cooperate with each other to achieve such an immune response: immunoglobulin (Ig), T-cell receptors (TCRs), and molecules encoded by genes of the major histocompatibility complex (MHC). Two of the three are present only in lymphocytes, while MHC molecules are present on the surface of most cells. As a rule, one lymphocyte expresses (i.e., has present on the surface) only one of the seemingly endless number of different Igs or TCRs in the system, while it may express several MHC genes, at least both alleles of the father and mother.

The bloodstream carries an antigen to a regional lymphoid organ such as a lymph node. Here the antigen is met by the cells of the immune system. At least three different types of cells take up the antigen when it enters a lymph node or the spleen: the follicular dendritic cell (FDC), the macrophages, and B lymphocytes. Two of them establish immediate contact: these are the macrophages and B lymphocytes.

The FDCs act as depots for antigens. From the FDC, two other types of cell can receive antigens for further processing. The antigen is "processed" in the macrophages by being degraded into fragments and then these are presented to T lymphocytes. The other cell type that takes up, processes, and presents antigen is the B cell. Approximately 1% of the B cells which bind a given antigen become stimulated to divide actively and secrete antibody. Consequently, simply binding antigen is not sufficient to trigger a B cell into antibody secretion. For some antigens, T cells or a T-cell-produced factor is required for B cell stimulation. Possibly, for other antigens, so-called T-independent antigens, T cells are not required.

An immediate humoral immune response results from the invasion of bacteria, parasites, and other antigens with repetitive antigenic determinants on their surface, without the help of T cells. Such antigens are called T-independent antigens. In contrast to macrophages, B cells have Ig molecules on their surface to which antigen can be specifically attracted. One B cell makes one molecular species of Ig, with one given binding specificity, and displays it in 10^4 to 10^5 copies on its surface membrane.

1.2.2 Clones

Proliferation and maturation of the B cells happen without the help of T cells, but the cooperation of macrophages in the form of cytokines, which help the B cells in their clonal expansion, is needed. This so-called T-cell-independent, B-cell stimulation is initiated and propagated by crosslinking of Ig in the surface membrane through repeated antigenic determinants. After an antigenic stimulation, B cells differentiate into a population (clone) of terminal cells. A clone is a family of cells which are all derived from one parent cell. These cells cannot divide further. Certain cells (plasma cells) produce antibodies with the same specificity as the immunoglobulin receptors which are present on the B cell surface. Plasma cells produce antibodies for several days at the approximate rate of about 2×10^3 molecules per second, and then die. Another group of B cells (memory cells) produce some antibodies, but they do so at a much slower rate than plasma cells.

1.2.3 Major histocompatibility complexes

Historically, our knowledge of cell-mediated immunity has developed from studies of the rejection reaction. If normal tissue (skin) is transplanted from one unrelated human to another, it is rejected by an immunological process. This occurs because the donor and host carry different alleles of genes as codes for the normal components of the cell surface, known as histocompatibility antigens, and these differences are recognized by the immune system. The tissue transferred from one individual to another would also be rejected because it has normal histocompatibility antigens: moreover, it may also have additional cell-surface components unique to the donor tissue. Differences in histocompatibility antigens are potent barriers to transplantation.

Histocompatibility antigens are classified on the basis of the strength of the rejection reaction they produce. One such locus which is found to be associated with a strong rejection reaction is called the major histocompatibility complex. The MHC appears to play a central role in cell–cell recognition. The genes of the MHC code can be grouped into two main classes. Class I products are glycoproteins which are expressed on the surfaces of all cells. Class II are expressed only on some of the cells actively involved in the immune system.

The process of antigen recognition is the interaction of cellular receptors and the antigenic determinants that structurally determine the specificity of antigen. If the structures of cellular receptors and antigenic determinant are complementary (as lock and key) then a complex "receptor–determinant" is formed. This complex can rest on a cell surface or depart from it. Consequently, antigen has been recognized. Then one of the immune reaction forms is developed. Since about 10^{10} lymphocytes circulate in the blood, the immune system has information about virtually all antigens that may penetrate the organism. Any such antigen has a finite probability of meeting a lymphocyte in the blood that recognizes it automatically. With a series of stimulations of differentiations and generation of antibody molecules, such recognition leads to the development of an immune response. For example, the immune reaction to pathogens of viral diseases (influenza, measles, poliomyelitis, viral hepatitis, etc.) includes both types of immune response: the humoral one and the cytotoxic T-cell response. Antibodies neutralize viruses circulating in the blood, while cytotoxic T lymphocytes (CTLs) recognize virus-specific antigens expressed on virally infected cells, and they destroy these cells.

During the immune response, the molecules and cells of the immune system interact with each other, as well as with other cells and molecules in the organism. Therefore, to model this process we can use the principles of cellular and molecular kinetics. We have now considered some of the main notions and phenomena of immunology. However, it will be necessary to introduce further information and phenomena as the discussion develops.

1.3   Cellular and Molecular Kinetics

As discussed above, cellular and molecular kinetics form the basis of the entire immune process. These processes can be quite well defined using conservation equations and chemical mass–action principles. In general, the cellular population (or concentration) x_i of the ith class may be described by

$$\frac{dx_i}{dt} = \text{source rate} - \text{death rate} + \text{division rate}$$
$$+ \text{ rate differentiation to} - \text{rate differentiation from , (1.2)}$$

or

$$\frac{dx_i}{dt} = v_i(t) - \frac{x_i}{\tau_i} + p_i(.)x_i + \sum_{j \neq i} 2p_j(.)p_{ji}(.)x_j$$
$$- \sum_{k \neq i} 2p_i(.)p_{ik}(.)x_i \ , \tag{1.3}$$

where $v_i(t)$ is the source term (from bone marrow via blood), τ_i is the death time constant, and $p_i(.)$ and $p_{ji}(.)$ are appropriate growth coefficients (including probabilities of stimulation and differentiation from one class to the other). These coefficients or probabilities represent parametric feedback control in the immune system which is of a very complex nature. Indeed, it is these terms upon which much immunological research is currently focused, i.e., in what manner is cell production activated and controlled by mainly molecular substances? Consequently, $p_i(.)$, $p_{ji}(.)$, and $p_{ik}(.)$ are functions of, primarily, molecular concentrations which are secreted by various classes of cells as naturally or artificially induced parametric controls. They may be deterministic functions or random processes depending on the approximation used. Here, i refers to different cell types: for example, it might refer to B-, T-, and macrophage-cell lineages, such as resting cells, excited cells, cytotoxic cells, suppressors, helpers, memory cells, and plasma cells (which generate antibodies, Ab). Also, other killer cells and mast cells could be included.

An mth class of molecular concentrations, y_m, may be described by

$$\frac{dy_m}{dt} = \text{molecular source rate} - \text{net death rate}$$
$$+ \text{dissociation rate of appropriate complexes}$$
$$- \text{association rate of appropriate complexes} \ . \tag{1.4}$$

This is usually approximated by

$$\frac{dy_m}{dt} = w_m(t) + \sum_i \beta_{im} x_i + \sum_{n \neq m} c_{mn} y_n - \sum_{l \neq m} c_{lm} y_l y_m - \frac{y_m}{\tau_m} \ , \tag{1.5}$$

or more accurately described by

$$\frac{dy_m}{dt} = w_m(t) + \sum_i \beta_{im}(.)x_i + \sum_{n \neq m} c_{mn}(.)y_n$$
$$- \sum_{l \neq m} c_{lm}(.)y_l y_m - \frac{y_m}{\tau_m} \ . \tag{1.6}$$

Here, $w_m(t)$ is an external source rate, τ_m is a lifetime, β_{im} is an ith-cell source rate for generating y_m, and c_{mn} and c_{lm} are appropriate coefficients of dissociation and association, respectively. In equation (1.5), these coefficients would be functions of the appropriate $y_k(t)$. The term y_n refers to immune system complexes of bound molecules which may dissociate. They may be assumed to be deterministic or stochastic, depending again on the approximation desired or the information available. The m refers to the mth class of molecules, such as Ab, Ag, the appropriate cell receptor, lymphokine, interferon (IFN), or other molecular substances. Note that the antigen–antibody model in equation (1.1) is a simple example.

If x_i^s and y_m^s refer to a particular compartment, s, or organ with migration between compartments, then equations (1.2)–(1.6) must include net migration terms: for example,

$$
\begin{aligned}
\frac{dx_i^s}{dt} = \; & v_i^s(t) - \frac{x_i^s}{\tau_i^s} + p_i^s(.)x_i^s + \sum_{j \neq i} 2p_j^s(.)p_{ji}^s(.)x_j^s \\
& - \sum_{k \neq i} 2p_i^s(.)p_{ik}^s(.)x_i^s + \sum_{u \neq s} \delta_{i,s,u}(.)x_i^s - \sum_{r \neq s} \delta_{i,r,s}(.)x_i^r \; . \quad (1.7)
\end{aligned}
$$

In this equation, superscript s shows possible dependence on the compartment or organ. For example, an inflamed spleen should cause more stimulation of appropriate cells. In general, the migration coefficients, $\delta_{i,s,u}(.)$ and $\delta_{i,r,s}(.)$, could be deterministic or stochastic functions of appropriate $y_m(t)$ terms since certain lymphokines (e.g., macrophage migration inhibition factor, MIF) manipulate migration coefficients. A similar molecular version of equation (1.5) may be developed.

If the appropriate $c(.)$, $p(.)$, and $\beta(.)$ terms are collected to form parametric feedback control terms (via composite molecular terms in general) and $v_i(.)$ and $w_m(.)$ are additive (linear) control terms, the immune system's components are synthesized as bilinear control systems. The system in totality is synthesized as bilinear systems (BLSs) which are coupled together by non-dynamical, non-linear terms. Typically, the additive (linear) controls are synthesized by appropriate antigen inoculation, cellular transplant (of bone-marrow cells), or specifically functioning cells, and the parametric (bilinear) controls are synthesized by interleukins, interferons, etc. Non-linearities in the feedback involve saturation as appropriate for limited molecular uptake in chemical mass–action balances, as well as killing terms such as appropriate molecules of cancer cells (studied in Part III).

If the state vector $x(t)$ is composed of all cell concentrations, $x_i(t)$, and all molecular concentrations, $y_m(t)$, then the BLS representation of equations (1.3)–(1.6) becomes

$$\frac{d}{dt}x = Ax + \sum_k B_k u_k x + Cv \qquad (1.8)$$

where A, B_k, and C are appropriate constant matrices, u_k is an appropriate control term (actually feedback functions of certain molecular $y_m(t)$ terms), and v is the term for appropriate additive controls (e.g., bone-marrow cells). While the BLS is a valid open-loop representation, the corresponding closed-loop system in which all non-linear feedback variables $(y_m(t))$ are measured becomes conditionally linear, with a corresponding conditional state-transition matrix.

While the composite immune system as modeled here is highly non-linear, its subsystem control can be studied in terms of bilinear system control. Furthermore, by appropriate measurement of the molecular concentrations which appear to be non-linear, the composite system ideally becomes conditionally linear, i.e., linear in the non-measured states given that the non-linear ones are measured (Mohler, 1991). Finite dimensional state estimators and other properties of such systems may then be utilized (Kolodziej and Mohler, 1986).

While the dynamical model structure of the mathematical model is largely formulated by the above, there is a great deal of uncertainty in the specific non-linear coupling functions. Indeed, this is the focus of most of experimental immunology. Fortunately, it seems that the components of the dynamical immune system all follow the above coupled bilinear structure as a consequence of cellular division and chemical mass–action dynamics. However, the precise stimulation, such as the results of particular interactions between macrophages, antibodies, T-helper cells, T-suppressor cells, interferon, and interleukins is not well established for control of a particular disease.

Models with the number of state variables of the order of at least 10 or 20 may be required in the long term for an effective analysis of such diseases as AIDS. Simple, low-order approximations may be analyzed with the methodologies available for computational convenience, and since there is insufficient data in most cases. Some of these are studied in later chapters.

Chapter 2

Model of Infectious Disease

In this chapter we study the mathematical model of disease which was proposed by Marchuk and studied by Asachenkov (Asachenkov, 1982) and Belykh (Belykh, 1988; Belykh and Marchuk, 1979; Belykh and Asachenkov, 1985). Our model is only a crude approximation and generally requires further refinement. However, even in this form it enables us to bring various factors essential to the dynamics of an infectious disease into a system. We believe our model is useful for studying the general picture of the course of a disease and clarifying some observational results. It is also possible that separate parts of the theory can be used for finding effective methods of treatment.

In this discussion, we start to construct the model on the basis of equilibrium relationships for each component of the immune system. From this perspective, particular characteristics of the functioning immune system are not essential for analysis of the disease dynamics, but, instead, the principal laws of the development of the organism's defensive reaction play a key role. Hence, while constructing the simplest model we will not distinguish between the cellular and humoral components of immunity antagonistic to the antigens which have penetrated the organism. We will assume only that the organism does have such defensive components at its disposal. We shall call them antibodies, regardless of whether we deal with the cellular or the humoral system of immunity. In the model, we also assume that the organism has sufficient macrophage resources utilizing the waste of the immune system's reaction, and also other non-specific factors needed for normal functioning of the immune system. In this respect, we confine ourselves to considering three components: antigen, antibody, and plasma cells for generated antibodies. We shall frequently refer to the stimulators of an infectious disease (antigens) as viruses, placing no precise biological meaning on this term. Therefore, in the model the virus is a multiplying pathogenic antigen. It should also be noted that during illness the degree of organ damage due to viral attack (by antigens) is of great significance, since it leads, in the final analysis, to lower activity of the immune system. This phenomenon should be accounted for in mathematical models.

Note that the simplest mathematical model in this interpretation permits distinct variations which can help us to find probable explanations of some important features of the operation of the immune system: the development of subclinical, acute, and chronic disease processes and their possible therapy.

2.1   Development of Marchuk's Model

The essence of the immune response to an invasion of genetically different substances (antigen), including disease stimulants, is to produce specific substances (antibody molecules, cell-killers) which are capable of neutralizing or destroying this antigen. In these terms, an infectious disease can be interpreted as a confrontation between the population of disease stimulants and the body's immune system. In this connection, as a first step, we distinguish the following main characteristics of disease:

- The concentration $V(t)$ of viruses (by viruses we mean multiplying pathogenic antigens).
- The concentration $F(t)$ of antibodies (by antibodies we mean substrates of the immune system, i.e., neutralizing viruses such as immunoglobulins and cell receptors).
- The concentration $C(t)$ of plasma cells. This is the population of carriers and producers of antibodies (including here immunocompetent cells and all immunoglobulin producers).
- The relative characteristic of a damaged organ $m(t)$.

It is a simple matter to derive the disease dynamics from conservation of matter principles, in a way similar to that in Chapter 1. Consequently, the model is a system of non-linear, ordinary differential equations:

$$\frac{dV}{dt} = (\beta - \gamma F)V$$

$$\frac{dC}{dt} = \xi(m)\alpha V(t - \tau)F(t - \tau) - \mu_c(C - C^*)$$

$$\frac{dF}{dt} = \rho C - (\mu_f + \eta\gamma V)F$$

$$\frac{dm}{dt} = \sigma V - \mu_m m \ . \tag{2.1}$$

Usually, for equations with delay the initial conditions are given on an interval $[t^0 - \tau, t^0]$. However, in the biological sense of the processes described, until the moment of infection (where $t = t^0$) there were no viruses in the organism: $V(t) \equiv 0$ for $t < t^0$, and therefore the initial conditions can be given at the point t^0. In what follows, when we discuss initial conditions for equations of this kind we mean $V(t) \equiv 0$ for $t < t^0$. We have:

$$\begin{aligned}
V(t) &= 0 \text{ for } t \in [-\tau, 0] \\
V(t^0) &= V^0 > 0, \quad C(t^0) = C^0 > 0 \\
F(t^0) &= F^0 > 0, \quad m(t^0) = 0 \ . \tag{2.2}
\end{aligned}$$

The first equation in (2.1) describes the change in the number of viruses in the organism. We assume that there is exponential growth of viruses with coefficient β. The term γFV represents the number of antigens neutralized by the antibodies denoted by F, and γ is the coefficient connected with the probability of neutralizing the viruses by an encounter with antibodies.

The second equation describes the growth of plasma cells. To this end, we take advantage of the simplest hypothesis on the formation of cascade populations of plasma cells. The immunocompetent B lymphocyte is stimulated by an antigen coupled with receptors of the T cell (VT-complex), and it initiates the cascade process of forming cells which synthesize the antibodies which neutralize antigens of this kind. Since, in our model, antibodies mean the substrates capable of binding with viruses (including the T-cell receptors), the number of lymphocytes stimulated in this way is assumed to be proportional to VF. Therefore, we arrive at the relationship describing the increment of plasma cells over a normal level, C^*, which is the constant level of plasma cells in a normal organism. The first term on the right-hand side of this second equation in (2.1) describes the generation of plasma cells; τ denotes the time during which a cascade of plasma cells is formed (for the development of the model without delay, see Appendix A); α denotes the coefficient which allows for the probability of an "antigen–antibody" encounter, the stimulation of the cascade reaction, and the number of newly generated cells. The second term on the right-hand side describes the fall in the number of plasma cells due to aging; μ_c is the coefficient equal to the inverse of the plasma cell's lifetime.

To obtain the third equation in (2.1), let us calculate the balance of the number of antibodies reacting with antigens. The first term ρC on the right-hand side describes the generation of antibodies by plasma cells: ρ denotes the rate of production of antibodies by one plasma cell. The second term, $\eta \gamma F V$, describes the fall in the number of antibodies due to binding with antigens: η denotes the number of antibodies needed for the neutralization of a single antigen. The third term, $\mu_f F$, describes the decrease in the antibody population due to aging, where μ_f is the coefficient inversely proportional to the time of decay of an antibody.

These equations do not account for the weakening of the vital activity of the organism during illness, which is caused by the fall in activity of the organs responsible for providing immunologic material – leukocytes, lymphocytes, antibodies, etc. – needed for the struggle with the multiplying viruses. Let us adopt the hypothesis that the productivity of such organs depends on the size of the damage to the target organ. To this end, let us consider an equation for the relative characteristic of the damage to the target organ. Let M be the characteristic of a normal organ (mass or area), and let M_0 be the corresponding characteristic of a normal part of the damaged organ. Then

$$m = 1 - \frac{M}{M_0}$$

represents the relative characteristic of damage to the target organ. For the intact organ it is zero, and for the completely damaged organ it is one. For this characteristic we consider the equation

$$\frac{dm}{dt} = \sigma V - \mu_m m \ ,$$

where the first term on the right-hand side (σV) is the degree of damage to the organ, and σ is a special constant for each particular disease. A decrease in this characteristic (m) is due to the recuperative capacity of the organism. The term $\mu_m m$ depends on m, with a proportionality coefficient μ_m, and characterizes the inverse of the recuperation period of the organ by e times (i.e., the organ-damage time constant).

It is clear that for severely damaged vital organs the productivity of antibody production drops. This drop is fatal for the organism and

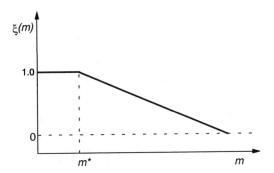

Figure 2.1. Decrease in antibody production for severely damaged organs.

leads to a lethal outcome. In our model, the damage factor of vital organs can be accounted for by the product $\alpha\xi(m)$. A typical graph for the function $\xi(m)$ is presented in *Figure 2.1*, where $\xi(m)$ over the interval $0 \leq m \leq m^*$ is equal to 1. This means that the efficiency of the immunologic organs in this interval does not depend on the severity of the illness. However, for $m^* \leq m \leq 1$ their productivity falls sharply, corresponding to the linear segment of the curve in this interval. The slope of this segment of the curve, as well as the quantity m^*, will be different for different diseases. The qualitative nature of this damage factor agrees with the observed patient response, in a crude sense.

We bear in mind that in our model we have a joint population of immunocompetent and antibody-producing cells, given by $C(t)$. In the absence of viruses in the organism, $C(t) = C^* > 0$; that is, C^* is, in fact, the normal level of immunocompetent cells in a normal organism. If such cells are absent, that is, $C^* = 0$, the organism is tolerant to a given antigen. However, it may turn out that the organism has no information about the given antigen, and consequently has no immunocompetent cells to oppose it. In such cases it is possible that the reaction involves immunocompetent cells with specific, similar receptors capable of stimulating an immune response to this antigen. We assume that the organism has a non-zero level of cells, C^*, with their own receptors, F^*, capable of causing an immune reaction. More refined trigger mechanisms of the immune system's reaction can be traced only with more complex mathematical models.

Note the following. First, by definition, $0 \leq m(t) \leq 1$, but in the model this fact is not explicit. We assume that if the organ is damaged completely, the viruses have nothing else to damage in this organ, i.e., $(dm/dt) = 0$ if $m(t) = 1(t > t^0)$. Second, let $t^0 = 0$ be an initial instant. Subsequently, we assume that variables of the model are continuous functions, the initial conditions given by equation (2.2) are non-negative, and all the parameters of the model are positive constants.

So, according to the model, the disease process is described as follows. At the moment of infection, $t^0 = 0$, a small population of viruses, V^0, penetrate into the body and begin to multiply and injure cells in the target organ. Some portion of the viruses bind with the receptors of the immunocompetent cells (with antibodies) and this leads to immune system stimulation resulting in the formation of a large population of plasma cells during the time period τ. These plasma cells begin to produce antibodies, which neutralize the virus population. The outcome of the disease is determined by the outcome of this competition. If the viruses can damage the organ severely during the formation of the immune response, the general condition of the organ deteriorates and, as a result, the response becomes less efficient. Antibody production declines and so does the probability of recovery.

2.2 Analytical Results

The validity of the following statements has been confirmed by Marchuk (1983) and Belykh (1988).

Statement 2.1
For all $t \geq 0$ there exists a unique solution of equation (2.1) with initial conditions as given in equation (2.2).

Statement 2.2
For all $t \geq 0$ the solutions of equation (2.1) with conditions as in equation (2.2) are non-negative.

Statements 2.1 and 2.2 demonstrate the features of the applicability of the model to real processes. For example, statement 2.1 guarantees that no case is biologically absurd, i.e., none of the model components has a finite escape time (i.e., reaches infinity during finite time). Statement 2.2 emphasizes the biological sense of model variables: they cannot have negative values.

Statement 2.3
The stationary solution

$$V_1 = 0, \quad F_1 = \rho C^*/\mu_f = F^*, \quad C_1 = C, \quad m_1 = 0 \tag{2.3}$$

is asymptotically stable if $\beta < \gamma F^*$. In this case, this is true at $F^0 = F^*$, $C^0 = C^*$, and $m^0 = m^*$ if the inequality

$$0 < V^0 < V^* = \frac{\mu_f(\gamma F^* - \beta)}{\beta \eta \gamma}$$

is valid, then $V(t) < V^0 e^{-at}$, where $a = \gamma \rho C^* / (\mu_f - \eta \gamma V^0) - \beta > 0$.

Statement 2.3 gives the condition of stability of the stationary solution in equation (2.3), which is interpreted as a healthy body state, and estimates the zone of its attraction in the case of infection of the healthy body. This estimate, V^*, has been called the immunological barrier against given types of viruses. If viruses cannot get over it ($V^0 < V^*$), no disease occurs since the virus population is removed from the body in the course of time. The model in equation (2.1) has another stationary solution which may be interpreted as a chronic form of disease.

Statement 2.4

The stationary solution

$$V_2 = \frac{\mu_c(\mu_f \beta - \gamma \rho C^*)}{\beta(\alpha p - \mu_c \eta \gamma)} > 0, \quad F_2 = \beta/\gamma$$

$$C_2 = \frac{\alpha \mu_f \beta - \eta \mu_c \gamma^2 C^*}{\gamma(\alpha p - \mu_c \eta \gamma)}, \quad m_2 = \sigma V_2 / \mu_m < m^* \qquad (2.4)$$

is asymptotically stable if

$$\mu_c \tau \leq 1, \quad 0 < \frac{f - d}{a - g\tau} < b - g - f\tau , \qquad (2.5)$$

where

$$\begin{aligned}
\alpha &= \mu_c + \mu_f + \eta \gamma V_2 \\
b &= \mu_c(\eta \gamma V_2 + \mu_f) - \eta \gamma \beta V_2 \\
d &= \mu_c \eta \gamma \beta V_2 \\
g &= \alpha \rho V_2 \\
f &= \beta \alpha \rho V_2 .
\end{aligned}$$

In the case where $\alpha \to \infty$, the second condition from equation (2.5) can be reduced to the inequality

$$0 < \beta - \gamma F^* < \left(\tau + \frac{1}{\mu_c + \mu_f} \right)^{-1} . \qquad (2.6)$$

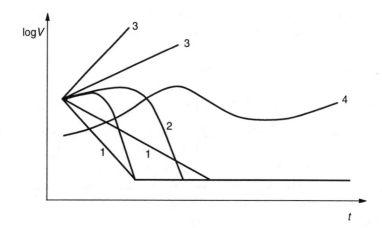

Figure 2.2. The possible forms of disease: 1 – subclinical; 2 – actual; 3 – lethal; and 4 – chronic.

Statement 2.4 defines the stability conditions for the stationary solution in equation (2.4), which is interpreted as a chronic, or latent, form of disease. It has been proved that, even in the case of a highly sensitive immune response ($\alpha \to \infty$), a stable, chronic form of disease is possible. Therefore, based on the mathematical model, we can establish the conditions for the development of the different forms of disease course. This is useful for the solution of practical problems.

2.3   Simulation

The infection of a healthy body by a small dose of viruses is simulated with appropriate initial conditions: $V(0) = V^0 > 0, F(0) = F^*$, $C(0) = C^*, m(0) = 0$. This simulation shows that there exist four, qualitatively different types of solution, which were interpreted as disease forms: subclinical, acute with recovery, chronic, and lethal outcome. They are illustrated in *Figure 2.2*.

A subclinical form of disease develops under the conditions of statement 2.3; it is characterized by a stable removal of the viruses from the body corresponding to those curves denoted by 1 in *Figure 2.2*. The acute form with recovery is a particularly dynamic type of virus behavior: first, there is a fast proliferation of viruses during several days and then a drastic contraction, practically to zero, due to a powerful immune response (curve 2). The chronic form of the disease is characterized by a persistent presence of viruses in the body; it arises in particular under the conditions of statement 2.4. The unlimited growth of viruses in the body and the damage of the entire organ are characteristic of a lethal outcome (curves denoted by 3). We now progress to a more detailed description of numerical experiments simulated for a particular form of disease, and also discuss their biological implications.

2.3.1 Subclinical forms of disease

The simulation results for the case when $\alpha < \gamma F^*$ are represented in *Figure 2.3*. Here, two situations are distinguished. *Figure 2.3(a)* shows the situation when $\alpha\rho < \mu_c\eta\gamma$ that corresponds to normal functioning of the immune response, and *Figure 2.3(b)* shows the situation when $\alpha\rho < \mu_c\eta\gamma$ that corresponds to a state of immunodeficiency.

For doses of infection smaller than the immunological barrier $(V^0 \leq V^*)$, the process of virus removal from the body depends neither on the infection nor the power of the immune response (curves 1 and 2). The elimination of viruses is possible due to the constant level of antibodies F^*. This situation seems to correspond to daily contact with small doses of antigen which penetrate into the body by respiration or with food.

With a higher dose of infection relative to the immunological barrier, the power of the immune response begins to play a major role. An efficient (normal) response is able to prevent an infectious disease [*Figure 2.3(a)*, curve 4]. With a weak immune response, the viruses penetrate through the immunological barrier $(V^0 > V^*)$, which leads to death [*Figure 2.3(b)*, curve 3]. Thus, the immunological resistance of people with a normal immune system $(\alpha\rho > \mu_c\eta\gamma)$ is much higher than in immunodeficient patients, who, naturally, are more susceptible to infection.

The case where $\beta < \gamma F^*$ can be interpreted as the vaccination of a healthy body by weakened, living antigens. Vaccination is meant to provoke a powerful immune response with the purpose of an essential accumulation of memory cells. According to our model, this is equivalent to an increasing level of immunocompetent cells, C^*, which are constantly present in the healthy body, and thus to a rising immunological barrier, V^*. The effect of vaccination is determined by the injected doses of antigen, as well as by the condition of the immune system. The simulation shows that injections of doses smaller than the immunological

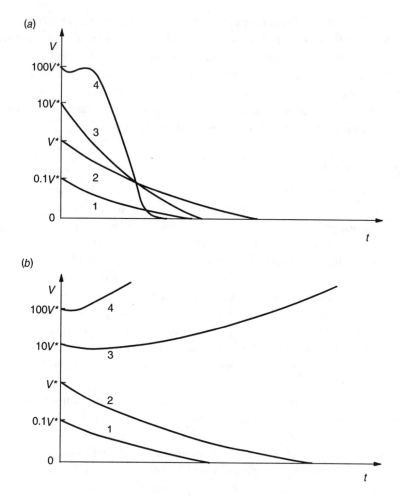

Figure 2.3. Subclinical forms of disease: (*a*) normal immune system, $\alpha\rho > \mu_c\eta\gamma$; (*b*) immunodeficiency, $\alpha\rho < \mu_c\eta\gamma$.

barrier have only a negligible effect because in this case the antigen is removed from the body, stimulating no immune response at all, or only a weak response. In neither situation is there an essential accumulation of memory cells. On the other hand, injections of larger doses ($V^0 > V^*$) into an organism with a normal immune system ($\alpha\rho > \mu_c\eta\gamma$) stimulate a strong immune response and lead to desirable results in treatment, whereas the vaccination of immunodeficient patients with high doses can cause a serious form of the disease [*Figure 2.3(b)*, curve 3].

2.3.2 Acute forms of disease

The simulation results of an acute form of disease with recovery in the case of a normal immune system ($\alpha\rho < \mu_c\eta\gamma$) are presented in *Figure 2.4(a)*. This form of disease occurs when $\beta > \gamma F^*$, and hence there is no immunological barrier to the stimulant of disease. As we noted earlier, the characteristics of this course of illness are: very fast (during several days) growth of viruses in the body until their number substantially exceeds the value of the infective dose, and very fast elimination of antigens from the body. The reason is two-fold:

- A high rate of virus multiplication leading to a fast accumulation of viruses in the body.
- The effective response of the immune system caused by the accumulated antigenic mass.

Figure 2.4(a) illustrates the course of an acute form of disease, depending on the rate of virus multiplication, γ, and the infective dose, V^0. The higher the multiplication rate at a given infective dose, the higher the maximum quantity of viruses, the faster they reach this number, and the faster the process terminates. This is explained by the fact that a high infective dose or a high rate of multiplication enables the viruses to reach those levels which effectively stimulate the immune system in a short period of time, and, as a result, the powerful immune system becomes capable of resisting the infection.

It appears that, if other circumstances are equal, the maximum level of viruses depends very little on (in fact is practically independent of) the infective dose [see *Figure 2.4(b)*]. We obtained an estimate of this maximum, $V_{\max}$, which is independent of V^0, which is given by

$$V_{\max} = \frac{(\beta - \gamma F^*)(\mu_f + \alpha)}{\gamma(\rho g - \eta\gamma f)} \; ,$$

where $f \in (F^*, \beta/\gamma), \alpha = \beta - \gamma f$, and $g = \alpha f e^{-\alpha\tau}(\mu_c + \alpha)^{-1}$. In *Figure 2.4(b)* we choose $f = [F^* + (\beta/\gamma)]/2$. Hence, in the case of the acute form of a disease, the value of the "peak of the disease" is independent of the infective dose, but determined by the immune characteristics of the organism with respect to viruses of a given type (the set of model parameters). The infective dose influences the moment of reaching the peak: the smaller V^0 is, the later the peak is reached.

Figure 2.4(c) demonstrates possible changes in the acute form of the disease with increasing coefficient of organ damage, σ. As a result of the

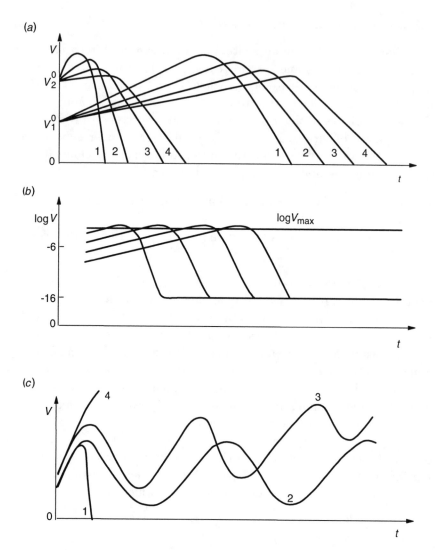

Figure 2.4. The acute form of disease: (a) dependencies on V^0 (for two different initial doses V_1 and V_2 and reproduction rate β (curve 1 is for high β, curve 4 for low β); (b) independence of maximal infection dose; (c) dependence on the organ damage σ.

organ damage, the acute form can turn into a chronic one (curve 2), into a chronic form with unpredictable outcome (curve 3), or a chronic form with fatal outcome (curve 4). The possibility of such a transition is due to the fact that, because of damage, the general condition of the body deteriorates, which makes the efficiency of the immune response decrease and the antibody production fall. Therefore, to prevent transition of the acute form into the more serious form, treatment should be aimed at either lowering the virus pathogeny or protecting the organ against damage.

2.3.3   Chronic forms of disease

We have already noted the possibility of the occurrence of a chronic form of disease from an acute infection, with serious organ damage [*Figure 2.4(c)*, curve 2]. Now we deal with other kinds of stable, chronic forms, which occur in particular under the conditions of statement 2.4. A characteristic of a typical chronic form is the flaccid dynamics of the viruses relative to the acute form. In this case, the passive virus dynamics lead in time to an equilibrium between new viruses and those neutralized by disease stimulants. Their concentration tends to a stationary level, V_2. With an increase in infective dose above V_2, the dynamics of the disease stimulants are more pronounced and transition into the acute form with recovery becomes possible [*Figure 2.5(a)*, curve 3]. The efficiency of the immune response can be enhanced by injecting higher infective doses. An analysis of the dependence between the course of the chronic form of disease and the infective dose of weak pathogenic viruses has brought us to the following conclusions:

- It is possible to treat a chronic form of disease by exacerbation (essentially increasing the number of viruses in the body).
- It is not reasonable for the immune system to react to small doses of virus in order to prevent a chronic form of disease.

A study of the dependence of the course of a chronic form of disease on the virus multiplication rate β [see *Figure 2.5(b)*] proves the existence of a stable, periodic solution (curve 2) and establishes that the treatment of the acute form using drugs, to decrease the multiplication rate, promotes the movement of the disease toward a chronic form. The existence and stability of the so-called hypertoxic chronic form of disease, when the damage level m_2 is higher than the limiting value m^*, is shown by Belykh.

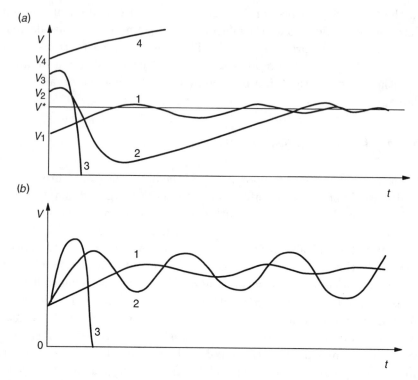

Figure 2.5. The chronic form of a disease: (*a*) dependence on infection dose V^0; (*b*) dependence on virus reproduction rate β, where $\beta_1 < \beta_2 < \beta_3$ (where β_1 is the virus reproduction rate for curve 1, etc.).

2.3.4 The origin of chronic forms and their possible treatment

A hypothesis on the immunological origin of different forms of disease has been proposed by Marchuk and Belykh: *chronic forms of disease are due to weak stimulation of the immune response.* This hypothesis is based on the following premises. In the framework of the model, the disease outcome (chronization or recovery) depends on the width of an interval (t_1, t_2) during which the concentration of viruses is decreasing, and consequently $dV/dt < 0$. If this interval is sufficiently wide (see *Figure 2.6*, the solid line), then the number of viruses decreases until it is close to zero, which is taken as recovery. Otherwise, i.e., in a

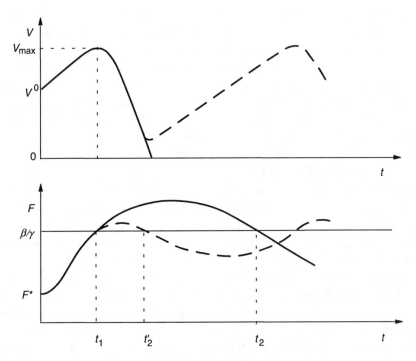

Figure 2.6. The occurrence of the chronic form of a disease from the acute form. The dynamics of the infectious process are depicted by a solid line. Biostimulators have been injected over a period of 23 days at an interval of $\Delta t = 1$.

sufficiently narrow interval (t_1, t_2), the number of viruses fails to approach zero, but when $t = t_2$ it reaches a minimum and begins to grow again for $t > t_2$. Then the process is repeated. This is the way the chronic form develops. Since $dV/dt < 0$ means $F(t) > \beta/\gamma$, the width of the interval (t_1, t_2) is determined by how long the latter inequality holds. Apparently, the more antibodies produced and the higher their maximum number, the wider the interval (t_1, t_2). If we allow for the fact that antibody production is essentially determined by the efficiency of the stimulation of the immune system, then the competence of the suggested hypothesis is obvious.

This hypothesis explains the results obtained from the simulation, namely, the transition of the acute form to the chronic one with serious organ damage, and the transition of the chronic form to the acute one

with high infective dose. In the first case, organ damage reduces the efficiency of immune system stimulation, which leads to a narrower interval
(t_1, t_2), and thus to a chronic disease process. In the second case there
is an inverse effect: a high dose of infection enhances the stimulation,
thus leading to a wider interval (t_1, t_2). It becomes theoretically obvious that the treatment of a chronic form of disease should promote a
widening of the interval (t_1, t_2). In practice, the stimulation of antibody
production (SAP) and disease exacerbation (biostimulation) have been
demonstrated by Belykh.

The action of the SAP factor is demonstrated by a three-fold increase in the quantity of antibodies when the SAP substance has been
injected at the peak of the immune response. This apparently makes the
interval (t_1, t_2) wider. The simulation of the action of the SAP-factor
shows its possible successful application in treating chronic forms of disease. In the case of exacerbation, the so-called biostimulation theory was
studied and its mathematical model was constructed. The basic notion
of this theory is the following. In a body which is subject to a stable,
chronic form of disease, a new, non-pathogenic, non-multiplying antigen
(biostimulator) is injected at some instant in time. The injections are
repeated over some discrete interval of time, and the dose of the injection grows with time. This leads to the situation in which, due to the
concurrence of macrophages between the two antigens (biostimulators
and disease stimulants), the immune response to viruses is blocked. So,
the immune system "forgets" the disease stimulants and this enables the
viruses to increase their antigenic mass. Some time later the biostimulator injections are terminated and then removed quickly from the body.
The organism is again face-to-face with the disease stimulants. However, the situation has essentially changed. During the interval when
the biostimulators were in the body, the amount of viruses in the body
reached the value which stimulates the immune system effectively. As
a result, a powerful immune response is formed: this leads to complete
elimination of viruses from the body and recovery follows.

The complex model corresponding to this theory has the form:

$$
\begin{aligned}
dV/dt &= (\beta - \gamma F)V \\
dF/dt &= \rho C - \eta\gamma FV - \mu_f F \\
dC/dt &= p_s(V)\xi(m)\alpha F_{(t-\tau)}V_{(t-\tau)} - \mu_c(C - C^*) \\
dm/dt &= \sigma V - \mu_m m \\
dB/dt &= -\gamma_B \phi B + f(t)
\end{aligned}
$$

$$d\phi/dt = \rho S - \eta_B \gamma_B \phi B - \mu_\phi \phi$$
$$dS/dt = p_s(B)\alpha_B \phi_{(t-\tau)} B_{(t-\tau)} - \mu_s(S - S^*) .$$

Here, in addition to previous notations which are relevant to the infectious process, new variables are introduced as follows: $B(t)$ is the concentration of biostimulators; $\phi(t)$ is the concentration of antibodies specific to the biostimulators; and $S(t)$ is the concentration of plasma cells producing the antibodies to act against the biostimulators. The function $f(t)$ describes the biostimulator injection and has the form

$$f(t) = \omega(n)\delta(t - n\Delta t) ,$$

where $\omega(n) = \omega_0 + \alpha n$, $\omega_0 > 0$, $\alpha > 0$, and $n = 1,\ldots,N$. The term N denotes the total numbers of injections and Δt is the time interval between consecutive injections. The functions $p_s(V)$ and $p_s(B)$ describe the probabilities of stimulating the immune system with the antigens V and biostimulators B, respectively. They have the form

$$p_s(V) = \frac{FV}{FV + \phi B}, \quad p_B = \frac{\phi B}{FV + \phi B} .$$

From inequality (2.6), we arrive at the conclusion that for successful treatment (complete elimination of viruses from the body) it is necessary to continue injecting biostimulators over a period of three to four weeks. *Figure 2.7* shows the simulation of the treatment of a typical chronic form of disease by exacerbation.

As can easily be seen, exacerbation results in considerable damage to the organ, and therefore biostimulators need to be combined with non-pathogenic therapy. Concerning biostimulation, which results in a powerful immune response, it is assumed that a sufficient number of memory cells have formed to create an immunological barrier to viruses of a given type. Then, subsequent contact between the body and the same viruses does not result in the infection of the organism. Other methods of treating the patient, for example by using antibiotics, even when the viruses have been completely eliminated from the body, do not provide an immunological barrier and subsequent contact with these viruses leads to the occurrence of a chronic form of the disease.

Elaborating on the idea of weak stimulation of the immune system, we point out that the failure of T–B cooperation is a possibility. The reason for such failure may be T-immunodeficiency, which may entail:

- The production of IgM antibodies only, rather than the chain IgM–IgG–IgA as usual.

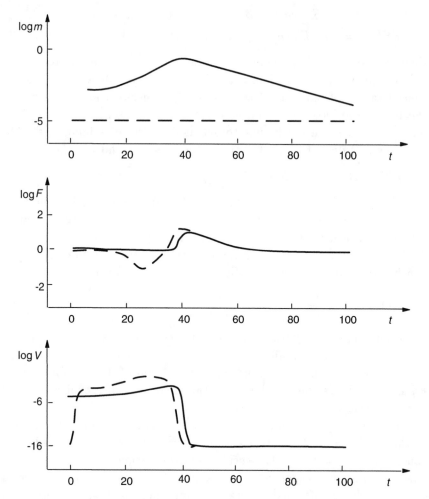

Figure 2.7. The treatment of the chronic form of a disease by aggravation.

- The failure of low-dose tolerance.

The first factor narrows the interval (t_1, t_2) (see *Figure 2.8*) and the second will force the immune system to react to small doses of viruses. This, in turn, may lead to chronization of the process. Thus, both low-dose tolerance and the chain of antibody production, IgM–IgG–IgA , are reasonable methods of preventing the development of chronic forms of disease.

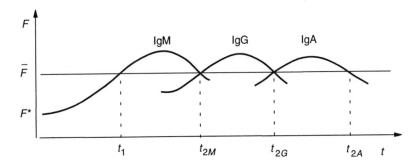

Figure 2.8. The chain production, IgM–IgG–IgA, which widens the interval (t_1, t_2).

Finally, this simple mathematical model of an infectious disease enables us to study the properties of the four main forms of disease. The conclusion is that the occurrence of any particular form of disease initiated by a small infective dose is not directly related to this dose. Rather, it is determined by the immunological status of the body with respect to a virus of a given type (the set of model parameters). This simple mathematical model opens up a new vista in science – the whole area of mathematical modeling of an infectious disease. The original model was modified in order to describe bi-infection, to investigate the role of the immunological memory (Skalko, 1983) and the organism temperature reaction (Asachenkov, 1982) in the pathological process, as well as to describe and interpret the experimental data on viral hepatitis (Nisevich *et al.*, 1984). Using this model, the practical immunological problem of the SAP action mechanism was solved (Stepanenko and Skalko, 1983). Moreover, new techniques for constructing methods which simplify the analysis of stationary-solution stability in time-delay systems, and their numerical integration, were obtained (Belykh, 1983).

Chapter 3

A Non-linear Synthesis of the Immune System

Obviously, the immunization process is extremely complex, and it would be impossible to develop an encompassing mathematical model to address all the relevant questions. However, immunological models and system studies can be useful for specific applications and as a basis for a general mathematical structure for future research. This joint analytical–experimental approach may lead eventually to a systematic understanding of the body's immune defense system and its stimulation for effective health care.

In general terms, the immune system is a communication, command, and control system to defend the body from alien intrusion and infection. Various immune subsystems, including the complement, humoral, and cell-mediated systems, are activated and deactivated according to the level of antigens and their chemical structure, in concert with other substances such as antibodies, suppressors, helpers, lymphokines, etc. Experimentation shows that the decay rate of bacteria is proportional to the product of its concentration with that of antibody and that of complement (prior to saturation), i.e., parametric control (Roitt, 1974). If the effect of antibody and complement were linearly additive, as for conventional linear systems, the bacterial decay would not be so rapid since superposition would apply. Such multiplicative effects (rather than superposition) seem very prevalent throughout immunology, as is shown in Chapter 1 and later in this chapter. A systematic overview of the immune system is presented by Mohler *et al.* (1980), and its bilinear synthesis (as well as bilinear control in general) is studied in Chapter 1 and by Mohler (1991).

3.1   The Humoral, Clonal-selection Model

The first mathematical clonal-selection model was presented by Bell (1970, 1972). It was presumed in this model that specific antigen-sensitive cells have on their surface antibody-like receptors that are capable of binding the antigen molecules to cells. Such bonds give antigenic stimulus, which controls the mitotic activity of the cells. The larger the numbers of receptors that are bound to an antigen, the higher the antigenic stimulus. There are four major types of B cells considered here: these are target cells (small lymphocytes), proliferating cells, plasma cells, and memory cells. The target cells pre-exist in the body before antigen injection. When sufficient antigenic stimulus occurs they transform into proliferating cells. Such proliferating cells divide (under antigenic stimulus) and produce antibodies. When the antigenic stimulus decreases they transform in a non-symmetric manner into plasma and memory cells. They cannot divide and their main function is antibody production. Memory cells are regarded as functionally identical with target cells. Bonds between antigen molecules and both receptors and antibodies are considered reversible. They are characterized by an intrinsic association constant and are described by mass–action laws and pseudostationary state hypotheses. Further development and generalization of the humoral process produces the following mathematical model which is a BLS [see the general forms of equation (1.8)]:

$$\frac{dx_1}{dt} = \alpha u_1 x_1 - \frac{x_1}{\tau_1} + v_1$$

$$\frac{dx_2}{dt} = 2\alpha u_2 x_1 - \frac{x_2}{\tau_2}$$

$$\frac{dx_3}{dt} = -cu_3 x_3 - \frac{x_2}{\tau_3} + \alpha' x_2 + cx_4 + \alpha'' x_1$$

$$\frac{dx_4}{dt} = cu_3 x_3 - \left(c + \frac{1}{\tau_4}\right) x_4$$

$$\frac{dx_5}{dt} = v_2 - \frac{x_5}{\tau_5} - Nc\left(u_3 x_3 - x_4\right) \ . \tag{3.1}$$

The constants included in the model have the following definitions. The term τ_1 is the average lifetime or natural decay for the population

$x_i(t)$, where $i = 1, 2, 3, 4, 5$. The term α is the ICC division rate; α' and α'' are the rates of antibody production by plasma and immunocompetent cells respectively; c and kc are the constants of dissociation and association of immune complexes, respectively; and N is the coefficient that describes the general effect of total antibody action on antigen.

In equation (3.1), the terms may be related directly to equations (1.2)–(1.6). So, $x_1(t)$ is the concentration of immunocompetent cells (ICC), which are the small lymphocytes with identical receptors of affinity k with respect to the antigen (these cells under antigenic stimulus can differentiate into plasma cells or memory cells that are considered equivalent to ICCs in the model); $x_2(t)$ is the concentration of plasma cells [$C(t)$ in Chapter 2]; $x_3(t)$ is the concentration of sites free of antibodies [$F(t)$ in Chapter 2]; $x_4(t)$ is the concentration of immune complexes (bound antigen–antibody complexes); and $x_5(t)$ is the concentration of free, monovalent antigen [$V(t)$ in Chapter 2]. The variables of additive (v_1, v_2) and multiplicative (u_1, u_2, u_3) control are incorporated in the model. Additive variables describe: v_1, the rate of origination of new ICCs from bone marrow (it is generally considered as a constant value different for different values of affinity k); and v_2, the rate of antigen injection (it depends on some prescribed time policy).

Here, the multiplicative variables have the form

$$u_1 = p_s(1 - 2p_d), \quad u_2 = p_s p_d, \quad u_3 = kx_5 \tag{3.2}$$

where p_s and p_d are dependent on the affinity k and the probabilities of the following two events concerning antigens. The first is that antigens induce ICC stimulation, and the second event is that an ICC differentiates into a plasma cell. Formulae for calculating these probabilities were derived in the following way. Let $R(k, t)$ and $L(k, t)$ be the amounts of free and bound receptors of cells, respectively, per unit volume. According to the mass–action law and chemical equilibrium, $L = kRx_5$. Then the probability that a k-affinity receptor is occupied (bound) is equal to

$$p = L/(R + L) = kx_5/(1 + kx_5) \ .$$

The probability that n receptors from the total number m are bound is defined by Bernoulli's law

$$p^{(n)} = C_n^m p^n (1 - p)^{n-m} \ .$$

It is supposed that stimulation takes place if the fraction of bound receptors belongs to some interval $[\sigma_1, \sigma_2]$. Then

$$p_s = \sum_{n=n_1}^{n_2} p^{(n)} = \sum_{n=n_1}^{n_2} C_n^m \frac{(kx_5)^n}{(1+kx_5)^m} \ , \tag{3.3}$$

where n_1 is the smallest integer that satisfies the inequality $n_1 \geq m\sigma_1$, and n_2 is the largest integer that satisfies the inequality $n_2 \leq m\sigma_2$. Since m is large, the distribution for p_s is approximated by

$$p_s = \begin{cases} 1 & \text{if} \quad \dfrac{\sigma_1}{1-\sigma_1} \leq kx_5 \leq \dfrac{\sigma_2}{1-\sigma_2} \\ 0 & \text{otherwise} \ . \end{cases} \tag{3.4}$$

The probability that a cell with affinity k and n bound receptors differentiates into a plasma cell is assumed to be equal to n/m. Then the probability that plasma cells are formed is

$$p_d = \sum_{n=0}^{m} \frac{n}{m} \ p^{(n)} = \frac{kx_5}{1+kx_5} \ . \tag{3.5}$$

The model describes the following picture of the humoral immune process. Before antigen injection $(t < t_0)$, there exists in the body the initial population of ICCs given by $x_1(t) = x^{*1} = u_1\tau_1 = constant$, with affinity k for antigen. Then, at $t = t_0$, at a low value of $u_2(t)$, an injection of non-multiplying, monovalent antigen takes place. The antigen begins to bind cell receptors. If the fraction of bound receptors has a small or large value, no ICC stimulation occurs because $p_s \approx 0$. As a result of this, no immune response (antibody production) arises. If the fraction of bound receptors belongs to some "average" interval $[\sigma_1, \sigma_2]$, then stimulation of the ICC population occurs. The stimulated cell divides, differentiates, and leaves the ICC population. It may transform into two plasma cells, with probability p_d, or into two memory cells, with probability $(1 - p_d)$, that are identical with their precursors (model assumption), and therefore join the ICC population. This process is described by the first term in the first equation in (3.1) for ICCs. Indeed,

$$\alpha\nu_1 x_1 = \alpha p_s(1 - 2p_d)x_1 = \alpha[2(1 - p_d)p_s x_1 - p_s x_1] \ .$$

Here, $p_s x_1$ is the number of stimulated cells that divide with rate α and leave the ICC population and $2(1 - p_d)p_s x_1$ is the number of memory cells that arise due to division. Then, $2\alpha p_d p_s x_1 = 2\alpha\nu_2 x_1$ plasma cells arise at division. Plasma cells, as well as ICCs, produce antibodies, but at a greater rate ($\alpha' > \alpha''$). Antibodies bind antigen with an association constant k_c: this leads to the formation of the immune complex. These, in turn, can dissociate with constant c and therefore be a source of free antibodies and antigen. If the amount of antigen is not increased, then, due to the neutralization of antigens by antibodies and the natural decay of the antigens, the ICC-stimulation decreases, which leads to a decrease in antibody production. Here, $p_d \to 0$, as antigenic stimulus disappears the accumulated plasma cells, antibodies, and immune complexes are only removed from the body by natural decay. It is assumed that memory cells and ICCs have long lifetimes in comparison with other populations. Hence, their decay is slow and this results in immunological memory.

It should be noted that the humoral immune system involves relatively fast and relatively slow modes of response, creating a very stiff process with interesting numerical integration problems for computer simulation if association and dissociation are considered to occur as in equations (3.1)–(3.5). Conventional Runge–Kutta integration algorithms were not found effective unless instant equilibrium is assumed to occur in association and dissociation. An adaptive Gear algorithm, however, was found to be very effective.

3.2   A Cell-mediated Model

While the above analysis is basic to the immune process, most antigens (and possibly all) induce some stimulation of macrophages and T cells (i.e., thymus-processed stem cells), which in turn secrete molecules to act as other parametric controls on the B-cell response in antibodies. It is commonly accepted that several different subclasses of T cells play important functional roles in the control of antibody response. Many details of T–B interactions, however, are still not well defined. This is a very active area of experimental research. Still, a systematic analysis of this topic may help lead to a better understanding and could help focus experimental research. Generally, T-cell subclasses include short-lived, non-recirculating T_1 cells and long-lived, recirculating T_2 cells, the latter are precursors to helpers, T_H, to suppressors, T_s, and to cytotoxic killers, T_c. The B-cell generation of conventional antibodies, in turn, is stimulated by T_H and suppressed by T_s. It is assumed that molecules, sometimes called associative antibodies, which are specific T-helper factors, are generated by T_H.

According to the model formulated here (Mohler *et al.*, 1978), immune system T complexes of molecules and antigen are released from their T-cell binding via their F_c regions to a third-party cell such as a macrophage. In turn, macrophage-bound T complexes bind, through antigen, with B-cell receptors. Thus, they indirectly stimulate antibody production. Those T complexes which bind directly with B-cell receptors have almost negligible stimulatory effect, but do cause passive inhibition by blocking the B-cell receptors from further stimulation (Greaves *et al.*, 1973).

It is assumed here that macrophages are available for instant binding with the T complexes, but that they have a limited area (or receptor sites) available for such binding. This limitation results in inhibition by further direct blocking of B-cell receptors as noted above. While T molecules do lead to removal of antigen, it should be noted that all the evidence indicates that their major role is in the stimulation of B–Ab production or TCRs, which, in turn, eventually leads to antigen removal.

It is interesting that stimulation and passive inhibition arise naturally in this model via macrophage coupling and B-cell blocking, respectively. Thus, a separate definition of T cells is not required for these two functions. However, there seems to be experimental evidence for the existence of active suppressor T cells, T_s. Further assumptions implicitly assumed here include the neglect of spatial distribution and circulation times, any dependence of lifetimes on densities or concentration of cells or molecules, any coupling between cell receptors, and the role of the complement. While independent receptor sites are inherently assumed, it is easy to alter the model in order to account for multivalent binding and cross linking.

The B-cell component of the T–B cell model is essentially as given above in section 3.1. A major addition to this submodel includes specific stimulation and inhibition factors in the stimulation and differentiation functions p_s and p_d, and a great deal of research is in process which should eventually lead to a clearer understanding of this. Here, the development of these functions follows that of the above B-cell model, such that consideration of added T-complex concentration yields, $x_{4T}(t,k)$, for B-cell binding have the following heuristic description:

$$p_s \simeq \begin{cases} 1 & \text{for} \quad \gamma_1 \leq kx_5 + \gamma_T(x_{4T}/x_3) \leq \gamma_2, \quad x_3 > 0 \\ \\ 0 & \text{elsewhere} \end{cases} \tag{3.6}$$

and

$$
p_d \simeq \begin{cases} \dfrac{kx_5 + \gamma_T(x_{4T}/x_3)}{1 + kx_5 + \gamma_T(x_{4T}/x_3)} & \text{for } x_{4T} \leq x_{4M}, \quad x_3 > 0 \\[4mm] \dfrac{kx_5 + \gamma_T[2x_{4M} - (x_{4T}/x_3)]}{1 + kx_5 + \gamma_T(x_{4T}/x_3)} & \text{for } x_{4T} \geq x_{4M}, \quad x_3 > 0 \ . \end{cases} \tag{3.7}
$$

Here, k is the affinity of TCRs for antigen, γ_1 and γ_2 are limits of distribution of p_s, γ_T is a T–B coupling coefficient to account for T complexes lost to bridging and for increased stimulation of B-cell activity relative to that from simple antigen binding, and x_{4M} is the finite limitation of the number of bindings that can be made with available macrophage surface.

These equations are analogous to equations (3.4) and (3.5), except that kx_5 (which was equal to the ratio of bound to free receptor sites) has been replaced by the ratio of the number of receptors bound to antigens and TCR–antigen complexes divided by the number of free receptors. In addition, the key parameters γ_T and x_{4M} have been introduced. It is important that immunogenicity increases with the ratio of macrophage-bound T complexes to free antibody receptor sites, but that it decreases with this ratio after a certain level is reached.

It is interesting that the ensuing T-cell model is similar in mathematical structure to the above B-cell model for subclasses of cells. If T_1 is neglected (or averaged in with T_2 according to weighted parameter values), and macrophage dynamics is also neglected, the T-cell state may be defined by the following parameters: $x_{1T}(t, k)$ is the concentration of T_2; $x_{2T}(t, k)$ is the concentration of T_H; $x_{3T}(t, k)$ is the secreted molecular concentration; and $x_{4T}(t, k)$ is the concentration of T-complexes.

While the model is valid for all values of k, single, average values, in general, are again used, as for the discrete B-cell model in section 2.3. Consequently, k arguments may be dropped. In this formulation, active suppression is neglected (or assumed to be averaged in with the helper-cell mechanism). This is done because of the variety of coupling mechanisms which have been proposed for active suppression. Also, the T_H cells and the passive inhibition mechanisms do seem to make more sense from an immunological point of view. Consequently, the T-cell model is quite similar to the B-cell model with the BLS form

$$
\frac{dx_T}{dt} = A_T x_T + (b_{1T} u_{1T} + b_{2T} u_{2T}) x_{1T} + b_{3T} x_{3T} + c_{4T} \nu_{1T} \tag{3.8}
$$

where

$$A = \begin{bmatrix} -1/\tau_{1T} & 0 & 0 & 0 \\ 0 & -1/\tau_{2T} & 0 & 0 \\ \alpha'' & \alpha'_T & -1/\tau_{3T} & c' \\ 0 & 0 & 0 & -[c' + (1/\tau_4)] \end{bmatrix}$$

$$b_{1T} = \begin{bmatrix} \alpha_T \\ 0 \\ 0 \\ 0 \end{bmatrix}, \quad b_{2T} = \begin{bmatrix} 0 \\ 2\alpha_T \\ 0 \\ 0 \end{bmatrix}, \quad b_{3T} = \begin{bmatrix} 0 \\ 0 \\ -c \\ c \end{bmatrix}, \quad c_{4T} = \begin{bmatrix} 1 \\ 0 \\ 0 \\ 0 \end{bmatrix}.$$

Here, the T-cell parameters are defined in a similar way to the B-cell parameters:

$$\nu_{1T} = p_{sT}(1 - 2p_{dT})$$

$$u_{2T} = p_{sT}p_{dT}$$

$$p_{dT} \simeq \begin{cases} 1 & \text{for} \quad \gamma_{1T} \leq k' x_5 \leq \gamma_{2T} \\ 0 & \text{otherwise} \end{cases} \tag{3.9}$$

where γ_{1T} and γ_{2T} are appropriate limits of distribution.

If the T–Ab binding of antigen is negligible compared with conventional antibody binding of antigen, the free antigen concentration (x_5) remains unchanged, and the B-cell submodel is again defined by equations (3.1)–(3.5), but with equations (3.3) and (3.5) replaced by (3.6) and (3.7).

The time responses of significant state variables were simulated using suitable values for the parameters, assuming the antigen to be in a complete Freund adjuvant. It can be readily shown that the model does mimic reality in a general sense, and that the addition of T cells and M_ϕ (or macrophage-like cells) does enhance responsiveness in both a positive and a negative sense. The effect of T_s cells and their active suppressor factor might be introduced in a variety of ways, such as by negative-action bridges similar to the positive bridge from a T_H-generated antibody.

Further aspects not treated above include added, non-specific stimulation, the two-signal theory, and self/non-self discrimination. Actually, the assumption of an average affinity accommodates the first complication very nicely, although some different immune mechanisms may be involved. Also, it accommodates the paralysis caused by a single bond with one particular antigen, and the activation caused by a second bond, particularly through joint antigen recognition by B and T receptors. This two-signal theory has been established by Bretscher and Cohn (Bretscher and Cohn, 1970; Bretscher, 1975) and has been included in a simple T–B model by Comincioli *et al.* (1979). According to this model, six T–B equations of state of a form similar to the above are developed with positive and negative reactivity signals generated as some hypothetical functions of, respectively

- The ratio of receptors bound with antigen, and Ag–T complexes, to the total number of receptors.
- The ratio of receptors with a single antigen bond to the total number of receptors.

It seems reasonable in the two-signal theory that the bond which is between the carrier portion of the antigen and associative antibodies is effective for B-cell activation only if the associative antibody is first bound to macrophage-like cells (such as macrophages or reticular cells), rather than in free solution or on the T-cell surface. This agrees with the Feldmann theory used in the above model formulation.

An alternative theory is proposed by Nossal and Schrader (1975) who suggest that the second signal is provided by the binding of a soluble, non-specific entity (secreted by macrophages) to non-immuno-globulin receptors. They also hypothesize a third signal to be the release of T-cell mechanisms to regulate the ultimate stages, such as through p_d and antibody production. A similar three-signal theory is proposed by Schimpl and Wecker (1975), with the option that T-independent antigens can deliver the third signal directly to B cells.

Again, however, equations (3.1)–(3.9) form an appropriate fixed-affinity model, with p_s and p_d computed for each component. In general, of course, all cellular (including macrophages) and molecular reactions follow the basic conservation principles. Such dynamics will be discussed in some detail in subsequent chapters.

3.3 Preliminary Analysis

First the existence and uniqueness of a solution to equation (3.1) are examined. If $x_5(.)$ is non-negative, as required in the real world and by the model (shown below), then it is readily seen from equations (3.3) and (3.5) that $u_1(.)$ and $u_2(.)$ are uniformly bounded. Furthermore, they are seen to be Lipschitzian. Consequently, there exists in some interval $0 \leq t \leq t_1$ a unique solution to equation (3.1) with $x_5(.) \geq 0$.

However, as shown by Mohler and Hsu, global existence and uniqueness can now be established by showing uniform boundedness of the solution for any t_1. Positive invariance (i.e., $x(t) \geq 0$ if $x(0) \geq 0$ for all $t \geq 0$) is easily shown by using a convenient theorem developed by Takeuchi et al. (1978). For example, the system $dx/dt = f(x)$, $x(0) = x_0$ is positive invariant if: $x(t)$ is a unique solution on $[0, t_1]$, $t_1 < \infty$ and for any $i = 1, 2, \ldots, n$, if $x_i = 0$ and $x_j \geq 0 (j \neq i)$, then $f_i(x) \geq 0$. It appears that a similar sequence of arguments can be made to show unique, positive-invariant solutions to the cell-mediated immune model of equation (3.8).

It has been shown by Karanam et al. (1978) that Walsh functions may be used conveniently to identify BLS and more general, non-linear systems such as appear in immunology. This is due to the following properties of these functions. Walsh functions, $\psi_i(.)$, $i = 0, 1, 2, \ldots$, form an Abelian group under the rule of $\psi_m(.)\psi_n(.) = \psi_{m \oplus n}(.)$ where $\oplus$ represents no-carry modulo 2 addition. Also, integrals of Walsh functions are equal to an infinite sum of Walsh functions defined (Fine, 1969) as

$$\int_0^t \psi(x)dx = P\psi(t) ,$$

where $\psi(.)$ is an infinite-dimensional column vector of Walsh functions, and P is an infinite-dimensional operational matrix. However, finite-dimensional approximations with $m = 2^n$ Walsh functions, $m > 1$, yields

$$\int_0^t \psi_{(m)}(x)dx = P_{(m)}\psi_{(m)}(t) ,$$

where $P_{(m)}$ and $\psi_{(m)}$ are of the order m. In general,

$$P_{(m)} = \begin{pmatrix} P_{m/2} & (-1/2m)I_{(m/2)} \\ (1/2m)I_{(m/2)} & O_{(m/2)} \end{pmatrix} \tag{3.10}$$

where $m = 2^n$, $n = 1, 2, \ldots, P_{(1)} = 1/2$, and $I_{(m/2)}$ and $O_{(m/2)}$ denote identity and null matrices of order $m/2$, respectively.

The identification procedure consists of a representation of inputs, states, and any time-variant parameters by a finite sum of Walsh functions, which from the above properties permits an identification of parameters by simple algebraic methods. For purposes of illustration the method is applied to a simple immune equation, i.e., to identify the parameters α, τ_1, and τ_2 of the B-cell model in equation (3.1). In order to normalize time to $(0, 1)$, let $\tau = (t - t_i)/(t_f - t_i)$, where t_i and t_f

are the initial and final times of the data considered. Then the second component of (3.1) becomes

$$\frac{d}{d\tau}x_2 = (t_f - t_i)\left(-\frac{1}{\tau_2}x_2 + 2\alpha u_2 x_1\right), \quad 0 < \tau < 1 , \tag{3.11}$$

which may be approximated as

$$e^T \psi_{(m)} \simeq x_2(t_i) - \frac{(t_f - t_i)}{\tau_2}e^T P_{(m)}\psi_{(m)}(\tau)$$
$$+2\alpha(t_f - t_i)f^T \Lambda_z P_{(m)}\psi_{(m)}(\tau)$$

where z, e, and f are the Walsh coefficient vectors of $x_1(\tau)$, $x_2(\tau)$, and $u_2(\tau)$. Also,

$$\Lambda_z = [z, \Lambda_1^{(m)}z, \ldots, \Lambda_{(m-1)}^{m_z}]$$

and $\Lambda_i^{(m)}$, where $i = 0, 1, \ldots, [(m/2) - 1]$, are matrices of ones and zeros defined by

$$\Lambda_i^{(m)} = \begin{pmatrix} \Lambda_i^{m/2} & O_{(m/2)} \\ O_{(m/2)} & \Lambda_i^{(m/2)} \end{pmatrix} \quad \Lambda_{i+(m/2)}^{(m)} = \begin{pmatrix} O_{(m/2)} & \Lambda_i^{(m/2)} \\ \Lambda_i^{(m/2)} & O_{(m/2)} \end{pmatrix} ,$$

and $\Lambda_i^{(m/2)} = I_{(m/2)}$.

With $t_i = 160h$ and $t_f = 240h$ in equation (3.11), and using $m = 8$ Walsh function approximations, four equations are generated to solve the equation for the two parameters τ_2 and α. The following values are obtained: $\alpha = 0.0572$ and $\tau_2 = 20.7$, compared with actual values of 0.0578 and 21 in the simulation. To estimate τ_1, advantage is taken of the fact that $u_1 = 0$ for $t > 240h$ and $v_1 = \beta_k$, a constant source. This technique with normalization yields

$$a^T \psi_{(m)}(\tau) \simeq x_1(t_i) - \frac{(t_f - t_i)}{\tau_1}a^T P_{(m)}\psi_{(m)}(\tau)$$
$$+\beta_k p_1^T \psi_{(m)}(\tau) ,$$

where p_1^T is the first row vector of the matrix $p_{(m)}$, and a is the new Walsh coefficient vector of $x_1(\tau)$. With $t_1 = 300h$ and $t_f = 380h$, the value of τ_1 is 7.15×10^3. That used in the simulation was 7.14×10^3.

For analyzing reachability properties the above immune model was approximated by a simpler BLS with scalar input of the form

$$\frac{d}{dt}x_1 = \left[\alpha(2u - 1) - \frac{1}{\tau_1}\right]x_1$$

$$\frac{d}{dt}x_2 = 2\alpha(1-u)x_1 - \frac{1}{\tau_2}x_2$$

$$\frac{d}{dt}x_3 = \alpha''x_1 + \alpha'x_2$$

$$t \in [0,T], \; x_1(0) = x_{10} > 0$$

$$x_2(0) = 0, \; x_3(0) = 0 \; . \tag{3.12}$$

The terms $x_1(t), x_2(t)$, and $x_3(t)$, respectively, denote the population density of ICCs, plasma cells, and antibody. This model is deducible from the previous one, assuming that $p_s = 1$ and that the input $u(t)$ is the fraction of cells that remain ICCs (i.e., $0 \leq u(t) \leq 1$); $1 - u(t)$ is the fraction that differentiates into plasma cells (that is, $u = 1 - p_d$). Rate constants α, τ_2, α'', and α' are defined above; the stem source and antibody removal are neglected.

The theory of BLS reachability is presented by Mohler (1973) and Brockett (1975), where the latter results are obtained from Lie algebraic techniques. An immediate consequence of Brockett's reachability theorem is the following immediately applicable corollary.

Consider the BLS with scalar input $u(t)$, where $0 \leq u(t) \leq 1$:

$$\frac{d}{dt}x = Ax + u(t)Bx, \; x(0) \in \mathbb{R}^n_+ \; .$$

If the BLS is positively invariant, and $\text{Rank}(B) = 1$, then the reachable set $IB(t)$ is convex for all $t \geq 0$. Therefore, the immune system approximated by equation (3.12) can be regulated by varying the probability of differentiation p_d within a convex set of $\mathbb{R}^n$. It seems obvious that, for example, if the immune system is capable of producing two levels of antibodies, then it is theoretically possible to regulate the antibody production between these two levels. Other state variables can be similarly controlled. This analysis confirms the convexity of the problem of finding the reachable set $IB(t)$, which may be solved next. It is known from Hirschorn (1973) that the reachable set $IB(t)$ for equation (3.13) can be expressed in terms of e^{At} multiplied by a Lie-group exponential (Gillmore, 1974). Of course, the reachable set does have physical connotations relative to disease control which are important for future work.

The above models consider the generation of antibodies from plasma and immunocompetent cells. (Actually, the models assume that memory cells are included in the ICC population, the memory cells being the source of these antibodies.) Perelson *et al.* (1976) and Mohler *et al.* (1978) study the time-optimal control of antibody generation in terms of their ICC and plasma-cell source. Hsu derives a general, time-optimal control theorem for a class of BLSs which includes equation (3.12) and applies the theorem to this immunological problem. It is seen that the time-optimal strategy depends on antibodies generated from ICCs early in the response and those generated from plasma cells later, and normal antibody levels.

Chapter 4

A Model of Anti-viral Immune Response

One needs to bear in mind that the simple model of an infectious disease considered in Chapter 2 is a mathematical abstraction. The model does not describe a specific disease caused by a specific antigen. The main objective of our modeling was to describe general laws inherent in all infectious diseases. Clearly, the construction of models of a specific disease would demand a more detailed description of the process and appropriate clinical data for the identification of the model parameters. In any case, presumably, the laws derived within the framework of the simple model will remain unchanged. The model which will be studied in this chapter was proposed by Marchuk and Petrov (1983).

4.1 The Immuno-genetical Description of Anti-viral Immune Response

The immune response to stimulants of viral infections in an organism, such as influenza, measles, poliomyelitis, viral hepatitis, and others, includes two types of response: a "humoral" response, when the system of B lymphocytes produces antibodies; and a "cellular" response, when cytotoxic T-lymphocyte effectors accumulate in the organism. It is the cellular response that secures the defense of the organism. Antibodies neutralize viral particles circulating in the blood but they are not capable of freeing the organism from infection, since virions multiply inside the cells sensitive to a given antigen. Antibodies do not penetrate inside the cells. Cytotoxic-lymphocyte effectors which have accumulated after

49

the immune response detect cells affected by the virus and kill them. Here they act in their role as killers of the cells of the host organism. Thus, an anti-viral immune response of the cellular type seems to be of an auto-immune nature. This, however, is not to be confused with the real auto-immune reaction. The latter involves pathological reactions of the immune system against normal (unchanged) cells or normal, cellular antigenic substances. By their anti-viral immunity, lymphocyte-killers destroy the cells of the host organism affected by the virus. Apparently, this is the only way to clean the organism of viruses, provided, of course, that the intracellular defense mechanisms (interferon, enzymes which control the replication of nucleic acids, and others) of the cells affected by the virus have not successfully arrested their multiplication (Petrov, 1976, 1983).

Immunocompetent cells recognize antigens of freely circulating viral particles, and the accumulated T effectors, which are armed with a receptor apparatus against viral antigens, find the cells affected by the virus and, acting against the viral particle, kill the infected cells. Zinkernagel (1976), and other investigators after him, proved that it is the virus–transplantation antigen complex, rather than the viral antigen proper, that initiates the processes of recognition, activates the T-lymphocyte effectors into proliferation, and activates the cytotoxic effect of the accumulated T killers – the so-called double recognition. The main type of molecule which forms a complex with the virus is the transplantation antigen in the major histocompatibility complex (H–k and H–d in mice and HA–A, HA–B and HA–C in humans). The term "double recognition" emphasizes the fact that the structure of the recognition receptor of the lymphocytes is similar to the virus–transplantation antigen complex, instead of being like a viral particle.

Soon after Zinkernagel produced his work, B. Benaceraff and E. Unanue (1979) showed that the double recognition principle holds true for other kinds of antigens, not only viral, including antigens stimulating the humoral immune response (antibody genesis). In these cases, however, other antigens in the major histocompatibility complex (Ia in mice and, apparently, HA–D in humans) are the primary molecules complexing with the antigen. Macrophages are the main cells which present antigens to the T-lymphocyte helpers in the process of recognition.

Thus, gradually, the picture of the recognition process of viral and other antigens has become much clearer. Apparently, T lymphocytes recognize only the antigen histocompatibility complex, and a macrophage is the first cell to interact with foreign antigens.

Antigens of the major histocompatibility complex (Ia, H–2K, and H–2D) play a leading role in cell interactions. Using them, macrophages and lymphocytes recognize each other. If these antigens are not identical on the interacting cells (say they have genetically determined distinctions), the macrophage–T lymphocyte, macrophage–B lymphocyte, T–B, and T–T interactions do not occur.

The discovery of double recognition and the significance of the identity of the antigens of the major histocompatibility complex for the cooperation of cells interacting in an immune response has made the scenario of cellular events during response development more precise, without necessarily changing it, as we hasten to emphasize.

4.2 Marchuk–Petrov Model

Based on our current knowledge of viral infection dynamics, we can distinguish the following main model variables.

$V_f(t)$: concentration of "free" viruses (viral particles freely circulating in the body which are capable of multiplying in the cells of the organ sensitive to a given type of virus), equivalent to the term $x_5(t)$ in Chapter 3.

$M_V(t)$: concentration of stimulated macrophages (macrophages which interact with free viruses and express the complexes SD* and Ia* on their surfaces).

$H_E(t)$: concentration of T-lymphocyte helpers participating in a cell-mediated response.

$H_B(t)$: concentration of T-lymphocyte helpers taking part in the humoral response.

$E(t)$: concentration of T-cell effectors (killers).

$B(t)$: concentration of immunocompetent B lymphocytes capable of adopting the stimulation signal from stimulated macrophages, M_V, and helpers, H_B.

$P(t)$: concentration of plasma cells, equivalent to the term $x_2(t)$ in Chapter 3.

$F(t)$: concentration of antibodies.

$C_V(t)$: concentration of organ's cells infected by viruses.

The argument t (time) will be omitted in the following description for the sake of simplicity. Next, we make the following assumptions.

1. The quantities of "virgin" macrophages in the body (M) and those of the organ's cells (C) are considered constant and sufficiently large for the increase in the number of the stimulated macrophages, M_V, and of infected cells, C_V, to be proportional to the quantity of free viruses, V_f.

2. The adoption of a stimulation signal by lymphocytes leads, after a certain period of time which is necessary for the division and proliferation of cells, to the formation of the terminal cell clone. To stimulate helpers, a single signal from M_V is necessary, and a double signal (from M_V and a corresponding helper) is necessary to stimulate E and B cells.

3. Part of the terminal cell clone can be stimulated to form a new clone under the corresponding signal. The remaining part executes other immune functions such as helping in stimulation, killing cells, and antibody production.

4. The life cycle of the lymphocyte helpers H_E and H_B is over after the interaction with lymphocytes E and B, respectively.

5. For a certain period of time the infected cells, C_V, execute their normal functions. Their death is due either to the development of irreversible viral infection or to their elimination by effectors (E). The damaged mass of the organ is therefore the quantity of cells killed by viruses plus the quantity of cells killed by lymphocyte effectors. Severe damage of the organ worsens the general state of the body and therefore lessens the efficiency of the stimulation of the immune system.

Consequently, the classic conservation equations take the following form:

$$\frac{d}{dt}V_f = nb_E C_V E + pb_m C_V - \gamma_m M V_f - \gamma_f V_f F - k_V \sigma C V_f$$

$$\frac{d}{dt}M_V = \gamma_M M V_f - \alpha_M M_V$$

$$\frac{d}{dt}H_E = b_H \left[\xi(m)\rho_H M_V(t - \tau_H)H_E(t - \tau_H) - M_V H_E \right]$$
$$- b_P M_V H_E E + \alpha_H (H_E^* - H_E)$$

$$\frac{d}{dt}H_B = b_H^B \left[\xi(m)\rho_H^B M_V(t - \tau_H^B)H_B(t - \tau_H^B) - M_V H_B \right]$$
$$- b_P^B M_V H_B B + \alpha_H^B (H_B^* - H_B)$$

$$\frac{d}{dt}E = b_P\left[\xi(m)\rho_E M_V(t-\tau_E)H_E(t-\tau_E)E(t-\tau_E)\right.$$
$$\left. - M_V H_E E\right] - b_E C_V E + \alpha_E(E^* - E)$$

$$\frac{d}{dt}B = b_P^B\left[\xi(m)\rho_B M_V(t-\tau_B)H_B(t-\tau_B)B(t-\tau_B)\right.$$
$$\left. - M_V H_B B\right] + \alpha_B(B^* - B)$$

$$\frac{d}{dt}P = b_P^P\xi(m)\rho_B M_V(t-\tau_B)H_B(t-\tau_B)B(t-\tau_B)$$
$$- \alpha_P(P^* - P)$$

$$\frac{d}{dt}F = \rho_f P - \eta_f\gamma_f V_f F - \alpha_f F$$

$$\frac{d}{dt}C_V = \sigma C V_f - b_E C_V E - b_m C_V$$

$$\frac{d}{dt}m = \mu b_E C_V E + \eta b_m C_V - \lambda m \ . \tag{4.1}$$

The initial conditions must be added to this system. It can easily be seen that the stationary model solution corresponding to a healthy body state is the solution

$$V_f = M_V = C_V = m = 0$$
$$H_E = H_E^*, \ H_B = H_B^*, \ E = E^*$$
$$P = P^*, \ B = B^*$$
$$F = \rho_f\left(\frac{P}{\alpha_f}\right) = F^* \ . \tag{4.2}$$

As before (see Chapter 2), we are interested in the entire natural situation – infection of a healthy body by a small dose of free viruses, V^0. In connection with this, we consider that before the instant of infection, t^0, i.e., at $t < t^0$, the system is in a stationary state [equation (4.2)], but at $t = t^0$ infection by a small dose, $V(t^0) = V^0$, takes place. Other components at $t = t^0$ maintain their stationary values. Since our model is autonomous, we may take $t^0 = 0$ without loss of generality.

In the first equation in (4.1), dV_f/dt means the rate (speed) of change of the free virus population in the organism. The increase in this population is due to the appearance of viruses from the infected cells where they have been multiplying, i.e., due to the death of infected cells as a result of their lysis by effectors [the first term on the right-hand side of the first equation in (4.2)] or due to viral damage (second term). The positive constants n and p denote the average quantity of free viruses per infected cell death as a result of lysis by effectors and of viral damage, respectively. The constant b_E characterizes the degree of interaction between C_V and E. The constant b_m denotes the proportion of infected cell deaths due to viral damage. A decrease in V_f is, firstly, due to macrophage stimulation (third term on the right-hand side, the term M is a constant equivalent to the quantity of "virgin" macrophages in the body, γ_m is the coefficient of interaction between M and V_f). Secondly, the decrease is due to neutralization by antibodies (fourth term), and, thirdly, due to the infection of the organ's healthy cells (last term on the right-hand side, C is a constant denoting the quantity of healthy cells in the organ, σ is the characteristic of interaction between C and V_f, and k_V is the number of viruses which have penetrated into one healthy cell). Assumption 1 is used in this case.

Now we discuss the equation for the concentration of stimulated macrophages, M_V – the second equation in (4.1). The first term on the right-hand side describes the increase in macrophages due to the interaction between M and free viruses, V_f (assumption 1). The second term gives the natural decay in the number of macrophages due to aging. Here, $\gamma_M = \delta_{\gamma_m}$, $\delta \geq 1$, where δ is the reciprocal of the quantity of free viruses capable of interacting with one macrophage, and α_M is the reciprocal of the lifetime of stimulated macrophages, M_V.

The equations for the concentrations of helpers H_E and H_B have an identical structure [the third and fourth equations in (4.1)]. The first term on the right-hand side describes the increase in helpers due to the formation of the terminal cells' clone under the influence of the stimulation signal from M_V (assumption 2); the second term describes the decrease in the number of helpers due to a new stimulation (assumption 3); and the third term their decrease due to the formation of a stimulation signal for the effectors (E) and lymphocytes (B) (assumption 4). The final term describes their homeostasis in the body. Here the coefficients b_H, b_H^B, b_P, and b_P^B characterize the appropriate interactions; $\xi(m)$ describes the influence of the organ's damage on the stimulation of the immune system (assumption 5); ρ_H and ρ_H^B characterize the average number of cells formed in one clone; τ_H and τ_H^B denote the time for the formation of an appropriate clone; α_H and α_H^B are the reciprocals of the lifetimes of helpers H_E and H_B, respectively; and H_E^* and H_B^* are terms which denote constant levels of appropriate helpers in a healthy body.

The equation for the concentration of the lymphocyte effectors, E [the fifth equation in (4.1)], has analogous structure. However, it should be noted here that two signals are required for the stimulation of E: one from helpers, H_E, and one from stimulated macrophages, M_V. The third term describes the decrease in E due to lysis of infected cells C_V. The constants have the same meaning.

The conservation equation for antibodies, F [the eighth equation in (4.1)], has the same structure as in the simple model (see Chapter 2). The first term on the right-hand side describes antibody production by plasma cells; the second one, their decrease due to the neutralization of free viruses; and the third one, their natural decay.

Finally, we consider the equation for the organ's non-functioning part [the tenth equation in (4.1)]. The increase in m is due to the death of infected cells as a result of lysis by effectors and viral damage. The last term describes organ regeneration with rate λ.

4.3   Immune Barrier and Solution Comparison

The following are useful in the subsequent analysis (Belykh, 1988).

Proposition 4.1
For given non-negative initial conditions, a unique solution of the model exists for all $t \geq 0$.

Proposition 4.2
The non-negativity of the given initial conditions leads to the non-negativity of the model solution for all $t \geq 0$.

Corollary
Under the condition of proposition 4.2, for $P(0) = P^*$ and $V^0 > 0$, $P(t) \geq P^*$ for all $t \geq 0$: the equality holds only at $t \in [0, \tau_B]$.

Proposition 4.3
A sufficient condition for asymptotic stability of the stationary solution (4.2) is the inequality

$$(\gamma_m M + \gamma_f F^* + k_V C)(b_E E^* + b_m) > \sigma C(n b_E E^* + p b_m) \quad .(4.3)$$

These propositions serve to establish the features essential for any model of a real biological process. For example, one requirement is that the quantity of any process component does not become infinite in finite time; another is the non-negativity of model components, as the biology of the process dictates. Proposition 4.3 guarantees the existence of an immunological barrier, V^* (see Chapter 2), such that in a healthy body infected by a small viral dose, $V^0 < V^*$, the disease does not develop unless this barrier is exceeded ($V^0 > V^*$) or the stability conditions are violated.

On the whole, this system describes the same infections which we investigated before in Chapter 2: however, the description is more detailed from the immunological point of view. First of all, antigen exists here in two forms: free viruses (V_f), which are freely circulating in a body, and viruses (C_V) in the cells of an infected organ. The immune system also has two subsystems: a T system, which includes helpers for humoral (H_B) and cell-mediated (H_E) responses, and killers (E); and a B system which includes basic B cells (B) and plasma cells (P). Macrophages stimulated by free viruses (M_V) are the antigen-presenting cells. As before, F designates antibodies and m is some characteristic of organ damage. Thus, this model describes the cooperation of three cellular systems: T systems, B systems, and macrophages. For the stimulation of helpers, H_B and H_E, only one signal from macrophages M_V is required, while for the stimulation of B cells and effectors, E, two signals are required – from macrophages M_V and the corresponding helpers, H_B or H_E.

The immune system responds to antigen invasion in two ways. First is the humoral response through the production of antibodies, F, which neutralize only free viruses, V_f. Second is the cell-mediated response with an accumulation of effector or killer cells, E, which can destroy only the cells of the organ infected by viruses. Free viruses, V_f, originate from the death of infected cells, C_V, due to both irreversible processes inside the cell and the killing of infected cells by infectors, E. Now one can imagine the general picture described by the model.

The corresponding propositions for global existence, uniqueness, and non-negativity, as well as for the stability of a healthy state, were proved (Bocharov, 1983; Druzchenko, 1982). Numerical analysis of the model presented here [equation (4.1)] confirmed the main biological conclusions of the simple model. A new fact discovered when using this model was the possibility of a situation where very aggressive killers can in effect lead to the death of a body. The point is that killers destroy their own cells which are infected by viruses. On one hand this leads to organ damage, and on the other it facilitates the appearance of free viruses from the cell destroyed by killers. These "new" viruses begin to infect cells again and the situation is repeated. Such "chain reactions" can lead to the death of a body. It is possible to prevent this situation by suppression of the cell-mediated response.

Let us formulate proposition 4.4. in the following way.

Proposition 4.4
A sufficient condition for the stationary solution in equation (4.2) to be asymptotically stable is the validity of inequality $\beta^* < \gamma_f F^*$ where

$$\beta^* = \delta C \left(\frac{nb_E E^* + pb_m}{b_E E^* + b_m} - k_V \right) - \gamma_m M \ .$$

Stationary solution (4.2) corresponds to a healthy body state. The constant β^* is the effective rate of virus reproduction. As in the case of the simple model, we study the infection of a healthy body invaded by viruses. This means that the initial conditions for all variables at $t = t_0$ coincide with the stationary solution (4.2), except for those viruses for which $V_f(t^0) = V_f^0 > 0$. We assume that $t^0 = 0$. Using the property of non-negativity of the solutions, and from the right-hand sides of equation (4.1), as well as the method used for the analogous proof of the simple model (Chapter 2), the following propositions on the immunological barrier were proved (Belykh, 1988).

Proposition 4.5
Let $E = E^*$ and $dC_V/dt = 0$ at $t \geq 0$. If at $\beta^* < \gamma_f F^*$ the inequality

$$0 < V_f^0 < V^* = \frac{\alpha_f(\gamma_f F^* - \beta^*)}{\beta^* \eta_f \gamma_f} \tag{4.4}$$

is valid, then $V_f(t)$ decreases on the interval $[0, \infty)$, and for $t \geq 0$, $V_f(t) \leq V_f^0 e^{-at}$, where

$$a = \frac{\gamma_f \rho_f P^*}{\alpha_f + \eta_f \gamma_f V_f^0} - \beta^* > 0 \ .$$

Proposition 4.6
Consider dC_V/dt for $t \geq 0$. If at $\beta^{**} < \gamma_f F^*$, where $\beta^{**} = \delta C[\max (n, p) - k_V] - \gamma_m M$, the inequality

$$0 < V_f^0 < V^{**} = \frac{\alpha_f(\gamma_f F^* - \beta^{**})}{\beta^{**} \eta_f \gamma_f} \tag{4.5}$$

is valid, then $V_f(t)$ decreases on the interval $[0, \infty)$, and for $t \geq 0$, $V_f(t) \leq V_f^0 e^{-a_1 t}$, where

$$a_1 = \frac{\gamma_f \rho_f P^*}{\alpha_f + \eta_f \gamma_f V_f^0} - \beta^{**} > 0 \ .$$

Proposition 4.7

Let $|dC_V/dt| \leq (L-1)\delta C V_f$ at $t \geq 0$ where $L \geq 1$. If at $\beta^{***} < \gamma_f F^*$, where $\beta^{***} = \delta C[L \max(n, p) - k_V] - \gamma_m M$, the inequality

$$0 < V_f^0 < V^{***} = \frac{\alpha_f(\gamma_f F^* - \beta^{***})}{\beta^{***} \eta_f \gamma_f} \tag{4.6}$$

is valid, then $V_f(t)$ decreases on the interval $[0, \infty)$ and for $t \geq 0$, $V_f(t) \leq V_f^0 e^{-a_2 t}$, where

$$a_2 = \frac{\gamma_f \rho_f P^*}{\alpha_f + \eta_f \gamma_f V_f^0} - \beta^{***} > 0 \ .$$

Thus, these propositions give us, under different constraints, the estimates of immunological barriers V^*, V^{**}, and V^{***}. Recall that the immunological barrier means the infection dose which does not lead to the development of disease in a healthy body. In this case, the defense of the body is supplied by natural (normal) levels of immune components (E^*, B^*, C^*, etc.) and an added immune response is not needed. Hence we can consider, for example, killers $E(t)$ as slow variables in comparison with virus dynamics and assume $E(t) = E^*$ for $t \geq 0$. The constraint dC_V/dt means that we apply the hypothesis on quasi-stationary states (instantaneous equilibrium between interacting components). The constraint on $|dC^*/dt|$ is valid for $L = 2$ if $dC_V/dt > 0$ [see the right-hand side for dC_V/dt in equation (4.1) and keep in mind non-negativity of variables]. All of our constraints appear to be reasonable. We obtained estimates for the immunological barrier which are analogous to the barrier for the simple model (see Chapter 2). It should be noted that these estimates do not contain a constant connected with cellular response (at least for V^{**} and V^{***}). This outlines the main role of humoral components in defending a healthy body from small doses: it is important to destroy the infected cells using killers since such cells are the origin of the virus in the body.

Let $\xi(m) \equiv 1$ and $|dC_V| \leq (L-1)\delta C V_f$ at $t \geq 0$ where $L \geq 1$. Then, from the analysis of the right-hand side of equation (4.1), it follows that

$$\begin{aligned}
(\beta_{\min} - \gamma_f F) &\leq dV_f/dt \leq (\beta_{\max} - \gamma_f F) \\
\frac{dF}{dt} &= \rho_f P - \eta_f \gamma_f F - \alpha_f F \\
\frac{dP}{dt} &= \bar{\alpha} M_V H_B B|_{t-\tau_B} + \alpha_P(P^* - P) \\
\frac{dC_V}{dt} &\leq \delta C V_f - b_m C_V \ .
\end{aligned} \tag{4.7}$$

Here

$$\beta_{\max} = \delta C[L \max(n, p) - k_V] - \gamma_m M$$
$$\overline{\alpha} = B_P^{(P)} \rho_B$$
$$\beta_{\min} = \delta C[(L - 2) \min(n, p) - k] - \gamma_m M \ . \tag{4.8}$$

We can see that the right-hand side terms of equation (4.7) are similar to those of the simple model. The only difference is the difference between equations for plasma cells $P(t)$ and $C(t)$. However, if we simulate the real process (using experimental data), then the real plasma cell dynamics do not change and we can assume that $P(t) \approx CV(t)$. Under this condition it follows from equation (4.7) that the anti-viral model solutions $V_f(t)$, $F(t)$, and $C_V(t)$ are dominated by simple model solutions $V(t)$, $F(t)$, and $m(t)$ respectively. Therefore, all results connected with the simple model (see Chapter 2) must also be valid for Marchuk's model if the dynamics of plasma cells $P(t)$ and $C(t)$ coincide. It is clear now why the main regularities obtained in the simple model were confirmed by the anti-viral one. Based on these speculations, we put forward the following hypothesis, which is close to proposition 4.1.

Hypothesis
In the model given by equation (4.1) there exists a non-zero stationary solution corresponding to the chronic form of disease, the stability condition of which at sufficiently large α and $\alpha_p \tau B \le 1$ is

$$0 < \beta_{\max} - \gamma_f F^* < \left(\tau_B + \frac{1}{\alpha_p + \alpha_f}\right)^{-1} \ . \tag{4.9}$$

The following proposition concerning the immune barrier can be proved.

Proposition 4.8
For $t \ge 0$, $dH_V/dt = 0$, $H_\Sigma(t) \le H_\Sigma^{**} = $ constant, $L = \max (n, p) > k_V$, and at $t = 0$ let the initial conditions for the model in equation (4.1) be non-negative, $P(0) = P^*$, $F(0) = F^* = \rho_f P^*/\alpha_f$, and

$$0 < V_f(0) = V_f^0 < V_f^* = \frac{\alpha_f(\gamma_f F^* - \beta)}{\beta \eta_f \gamma_f}$$

where $0 < \beta = \delta H_{\Sigma}^{**}(L - k_V) - \gamma_m M < \gamma_f F^*$. Then the solution $V_f(t)$ decreases for $t \geq 0$ as $V_f(t) \to \infty$ when $t \to \infty$ and, moreover, $V_f(t) \leq V_f^0 e^{-at}$, where

$$a = \frac{\gamma_f \rho_f P^*}{\alpha_f + \eta_f \gamma_f V_f^0} - \beta > 0 \ .$$

The term V_f^* in this proposition is the immune barrier estimate.

Now it is useful to note a few points about such limitations. The constraint, $H_{\Sigma} \leq H_{\Sigma}^{**} = \text{constant}$, means that the amount of helpers in the body is limited and is biologically reasonable. Except for this, the model has a global solution. This means that none of its state variables can reach infinity in finite time. So for any finite-time interval, such a constant value of H_{Σ}^{**} exists, and from this point of view the constraint becomes mathematically reasonable. Since $H_{\Sigma} = H + H_1$, and in the absence of viruses the fraction of specific helpers is negligibly small (of the order of $10^{-6} - 10^{-7}$), we can assume that, even for an increase of 100–1000 times this amount during the immune response, $H_{\Sigma} = H$. From this it follows that at $t \geq 0$, $H(t) \leq H^* = (1 - k)k/\mu_H$ (the equality can take place only when $t = 0$), thus we can use the value H^* as an estimate for H_{Σ}^*. That is, $H_{\Sigma}(t) \approx H(t) < H^*$ for $t > 0$. Next, the constraint $L = \max(n, p) > k_V$ is biologically reasonable. Indeed, the term k_V is the average number of viruses which penetrate the healthy cell at the time of infection, but L is the average number of viruses realized after destruction of the infected cells. Since viruses reproduce inside the cell, the inequality $L > k_V$ is true.

Now, we discuss the result of proposition 4.8. First, the value of β given by $\beta = \delta H_{\Sigma}^{**}(L - k_V) - \gamma_m M$ can be considered as an upper estimate of some effective rate of multiplication of viruses. The term β is the maximum possible number of viruses arising in the body per unit of time. It describes the difference between virus reproduction (L) and infection (k_V), as well as their loss due to macrophage stimulation. Secondly, the estimate of the immune barrier, V^*, contains only the humoral component, in particular the natural level of antibodies, F^*. Cellular components are absent. This shows the leading role of the humoral response in prophylaxis of disease. Indeed, the increase in the natural level of antibodies, F^*, will be accompanied by an increase in the immune barrier V^*, and therefore this will decrease the probability of disease arising from contacts with infected patients. Thirdly, it can be shown that, with some constraints, the solutions of the Marchuk–Petrov model are dominated by solutions of the simple model. So these solutions can have the characteristics which were described by the simple model (as in Chapter 2).

4.4   Example: Viral Hepatitis

The Marchuk–Petrov model is applied for the investigation of various viral diseases such as influenza, measles, pneumonia, viral hepatitis, etc. (Marchuk and Berbentsova, 1989; Marchuk *et al.*, 1991; Nisevich *et al.*, 1984). Here, the application of this model to viral hepatitis is presented. For this purpose the model as given by equation (4.1) was modified in the following way (an explanation of terms is given in *Table 4.1*):

$$\frac{d}{dt}V_f = nb_{CE}C_V E + \nu C_V - \gamma_{VM}V_f - \gamma_{VM}V_f F$$
$$\qquad - \gamma_{VC}V_f(C^* - C_V - m)$$

$$\frac{d}{dt}M_V = \gamma_{MV}M^*V_f - \alpha_M M_V$$

$$\frac{d}{dt}H_E = b_H^E[\xi(m)\rho_H^E M_V(t - \tau_H^E)H_E(t - \tau_H^E) - M_V H_E]$$
$$\qquad - b_P^{H_E} M_V H_E E + \alpha_H^E(H_E^* - H_E)$$

$$\frac{d}{dt}H_B = b_H^B[\xi(m)\rho_H^B M_V(t - \tau_H^B)H_B(t - \tau_H^B) - M_V H_B]$$
$$\qquad - b_P^{H_B} M_V H_B B + \alpha_H^B(H_B^* - H_B)$$

$$\frac{d}{dt}E = b_P^E[\xi(m)\rho_E M_V(t - \tau_E)H_E(t - \tau_E)E(t - \tau_E)$$
$$\qquad - M_V H_E E] - b_{EC}C_V E + \alpha_E(E^* - E)$$

$$\frac{d}{dt}B = b_P^B[\xi(m)\rho_B M_V(t - \tau_B)H_B(t - \tau_B)B(t - \tau_B)$$
$$\qquad - M_V H_B B] + \alpha_B(B^* - B)$$

$$\frac{d}{dt}P = b_P^P \xi(m)\rho_P M_V(t - \tau_P)H_B(t - \tau_P)B(t - \tau_P)$$
$$\qquad - \alpha_P(P^* - P)$$

$$\frac{d}{dt}F = \rho_F P - \gamma_{FV}V_f F - \alpha_f F$$

$$\frac{d}{dt}C_V = \sigma C V_f(C^* - C_V - m) - b_{CE}C_V E - b_m C_V$$

$$\frac{d}{dt}m = b_{CE}C_V E + b_m C_V - \alpha_m \quad . \tag{4.10}$$

Table 4.1. Estimates of model parameters (from Marchuk et al., 1991).

Symbol	Physical meaning	Range of permissible values	Initial value (unit)
M^*	Concentration of class I- or class II- bearing macrophages (in lymph node, LN)	$5 \times 10^{-16} - 3 \times 10^{-15}$	10^{-15} (moles)
H_E^*	Concentration of hepatitis B antigen (HBAg)-specific helper cells cooperating with cytotoxic T cells (in LN)	$10^{-18} - 10^{-17}$	10^{-18} (moles)
H_B^*	Concentration of HBAg-specific helper cells cooperating with B cells (in LN)	$10^{-19} - 10^{-18}$	10^{-19} (moles)
E^*	Concentration of HBAg-specific cytotoxic T cells (precursors of T cells in LN)	$10^{18} - 10^{-17}$	10^{18} (moles)
B^*	Concentration of HBAg-specific B cells (in LN)	$5 \times 10^{19} - 5 \times 10^{18}$	10^{18} (moles)
P^*	Concentration of HBAg-specific plasma cells	$10^{-22} - 2 \times 10^{-21}$	4.3×10^{-22} (moles)
F^*	Concentration of HBAg-specific antibodies (in any compartment)	$2 \times 10^{-14} - 2 \times 10^{-13}$	8.3×10^{-14} (moles)
C^*	Concentration of hypothecates (in liver)	$5 \times 10^{-14} - 5 \times 10^{-13}$	5×10^{-13} (moles)
α_M	Rate constant of stimulated-state loss for macrophages	$1.0-1.5$	1.2 (day^{-1})
α_H^E	Rate constant of activated-state loss for H_E cells	$0.8-1.2$	1.0 (day^{-1})
α_H^B	Rate constant of activated-state loss for H_B cells	$0.8-1.2$	1.0 (day^{-1})
α_E	Rate constant of natural death for cytotoxic T cells	$0.33-0.5$	0.4 (day^{-1})
α_B	Rate constant of natural death for B cells	$0.05-0.1$	0.1 (day^{-1})
α_P	Rate constant of natural death for plasma cells	$0.33-0.5$	0.4 (day^{-1})
α_F	Rate constant of natural death for antibodies	0.043	0.043 (day^{-1})
τ_H^E	Duration of H_E cell single division	$0.4-0.8$	0.6 (day)
τ_H^B	Duration of H_B cell single division	$0.4-0.8$	0.6 (day)
τ_E	Duration of CTL division series	$2-3$	2.0 (day)
τ_B	Duration of B-cell division series	$2-3$	2.0 (day)
τ_P	Duration of B-cell division and differentiation resulting in the appearance of plasma cells	$3-4$	3.0 (day)

Symbol	Description	Range	Value
ρ_H^E, ρ_H^B	Number of helper T cells created by single division	2	2
ρ_E, ρ_B	Number of CTL and B cells in clone created by division series	4–32	16
ρ_P	Number of plasma cells in clone created by division series	2–5	3
ρ_F^*	Rate of immunoglobulin (IgG) production per plasma cell	$0.9 \times 10^8 - 2 \times 10^8$	1.7×10^8 (day^{-1})
b_H^E	Rate constant of H_E stimulation	$4 \times 10^{14} - 3 \times 10^{15}$	10^{15} (mol^{-1} day^{-1})
b_H^B	Rate constant of H_B stimulation	$4 \times 10^{14} - 3 \times 10^{15}$	10^{15} (mol^{-1} day^{-1})
b_P^E	Rate constant of CTL stimulation	$5 \times 10^{31} - 7 \times 10^{33}$	10^{32} (mol^{-2} day^{-1})
b_P^B	Rate constant of B-cell stimulation resulting in B-cell proliferation	$5 \times 10^{30} - 10^{33}$	10^{32} (mol^{-2} day^{-1})
b_P^B	Rate constant of B-cell stimulation resulting in B-cell proliferation and differentiation into plasma cells	$5 \times 10^{30} - 10^{33}$	10^{32} (mol^{-2} day^{-1})
γ_{MV}	Rate constant of macrophage stimulation	$10^8 - 4 \times 10^{13}$	10^8 (mol^{-1} day^{-1})
γ_{FV}	Rate constant of IgG binding to 22nm HBAg particles	$9 \times 10^{10} - 9 \times 10^{12}$	9×10^{11} (mol^{-1} day^{-1})
σ	Rate constant of hypothecates' infection with HBV	$2 \times 10^8 - 2 \times 10^9$	2.5×10^8 (mol^{-1} day^{-1})
b_{CE}	Rate constant of infected hypothecates damaged by CTL	$6 \times 10^{12} - 4 \times 10^{17}$	1.6×10^{15} (mol^{-1} day^{-1})
b_{EC}	Rate constant of CTL death due to lytic interactions with infected hypothecates	$6 \times 10^{11} - 4 \times 10^{16}$	1.6×10^{14} (mol^{-1} day^{-1})
b_m	Rate constant of infected hypothecates' damage due to HBV cytopathicity	$0.005–0.1$	0.01 (day^{-1})
α_m	Rate constant of hypothecates' regeneration	$0.1–0.3$	0.12 (day^{-1})
ν	Rate constant of secretion of HBAg and HBV particles per infected hepatocyte	$2 \times 10^2 - 10^5$	6×10^3 (day^{-1})
n	Number of HBAg particles released from infected hepatocyte damaged by CTL	$10^3 - 10^6$	2×10^4
γ_{VC}	Rate constant of HBV absorption by hypothecates	$2 \times 10^6 - 2 \times 10^9$	2×10^7 (mol^{-1} day^{-1})
γ_{VM}	Rate constant of non-specific and HBV removal	$0.1–0.7$	0.4 (day^{-1})
γ_{VF}	Rate constant of neutralization of HBAg and HBV particles by anti-HB IgG	$9 \times 10^{10} - 9 \times 10^{12}$	3×10^{11} (mol^{-1} day^{-1})
$b_P^{H_E}$	Parameter characterizing the suppression effect on the H_E population	$10^{-6} \times b_P^E$	—
$b_P^{H_B}$	Parameter characterizing the suppression effect on the H_B population	$10^{-4} \times b_P^B$	—

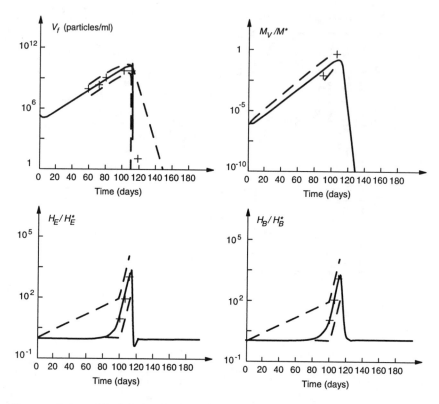

Figure 4.1a. Model simulating the course of acute viral hepatitis. The points marked "+" denote available experimental data; the boundaries of permissible values are marked by dotted lines; and the solid lines give the actual solution of the model.

Let us consider the data available concerning the dynamics of the suggested model variables. In the article by Marchuk *et al.* (1991) this problem is studied for acute hepatitis B (HBV). The clinical data available concern different aspects of HBV infection. In particular, the following characteristics have been investigated in detail for acute and chronic viral hepatitis: the dynamics of basic clinical tests; the variations in subpopulations of peripheral blood lymphocytes and macrophages; the level of steroid and protein hormones; the time-course of different HBV markers; HBV-specific antibodies; and morphological examinations of the liver. These data are summarized in *Table 4.1*.

The parameter values presented are based on explicitly introduced relations, including characteristic time-scales for the processes and the

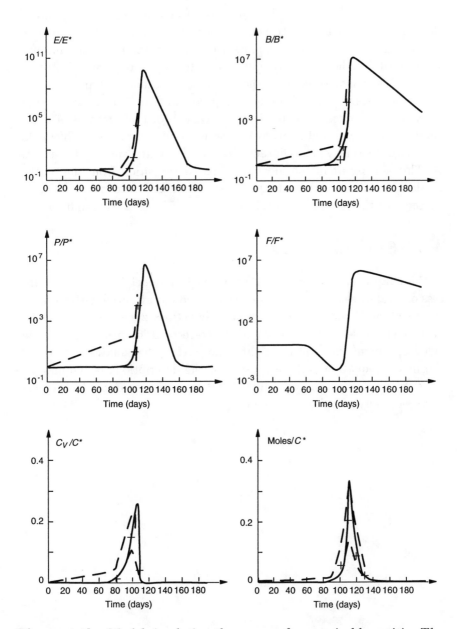

Figure 4.1b. Model simulating the course of acute viral hepatitis. The points marked "+" denote available experimental data; the boundaries of permissible values are marked by dotted lines; and the solid lines give the actual solution of the model.

characteristic amount of cells and substances considered in the model: these are thus open to criticism. Some of the parameters may be considered to be the intimate, quantitative characteristics of the immune system, while others are strongly related to the acute viral B hepatitis.

It is necessary to identify the mathematical model of anti-viral immune response which, according to experimental and clinical data, characterizes the course of acute viral hepatitis B for the purposes of mathematical modeling and analysis of a series of identification problems for various data sets which characterize the different aspects of hepatitis B. This will result in the most well-proved transition from the mathematical imitation to the actual disease course and outcome control. An example of the model simulation is given in *Figures 4.1(a)* and *(b)*.

4.5 Summary

In this chapter we constructed a mathematical model describing the process of viral disease. In this model we focused our attention on describing the reaction of the immune system to infectious agent invasion on the basis of clonal selection theory and the co-recognition principle. The delayed argument terms in the right-hand part of the equation describing the model [equation (4.1)] are used for the descriptions of lymphocyte division, multiplication, and differentiation processes into effector cells. Next we study lymphocyte traffic.

Chapter 5

Compartmental Models

As was mentioned above, the immune system includes a set of lymphoid organs. In this chapter we study the organ-distributed, immune dynamics. The first model demonstrates the stimulated humoral response as distributed in four major compartments. The second model considers the dynamics of recirculating lymphocytes throughout 12 compartments for which data are available for female rats (see *Figure 5.1*). Computer-simulated responses from these models are compared with available data. Unfortunately, the data are very difficult to obtain and by no means sufficient to adequately estimate model parameters. In many cases, the parameter values are merely reasonable guesses.

Obviously, such models can lead to a better understanding of immunology, and in the long term they can lead to more effective disease treatment. A maldistribution of immune mechanisms is a symptom of, for example, Hodgkin's disease, hepatitis B, psoriasis, and chronic lymphocyte leukemia. Also, one proposed method of localized tumor treatment is to tag appropriate migrating lymphocytes with a tumor-killing agent.

5.1   Stimulated Humoral Response

Here, a mathematical model is presented to show the distribution of the control function through several important, peripheral lymphoid organs as transfer media. Complementary to the models cited in Chapters 2–4, which consider cellular interaction and/or antigen–antibody reaction as a whole, this model is used to study the migratory pattern of immune constituents among organs after antigen stimulation. The qualitative

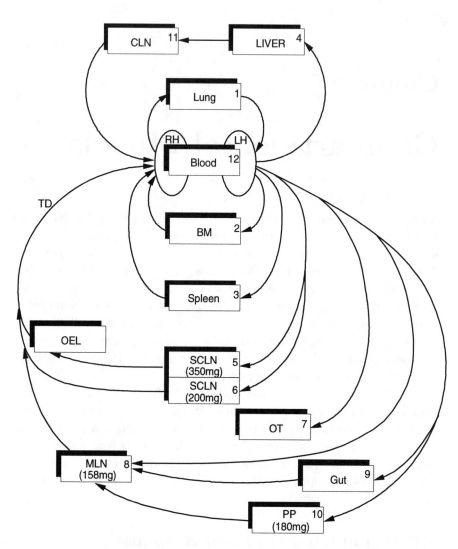

Figure 5.1. Lymphocyte traffic. Key: BM – bone marrow; SCLN – subcutaneous lymph node; OT – other tissue; MLN – mesenteric lymph node; OEL – other efferent lymphatics; PP – Peyers patches; CLN – coeliac lymph node; TD – thoracic duct; RH – right-hand side of heart; LH – left-hand side of heart. (The numbers correspond to the numbers in *Figure 5.3*.)

aspects of lymphocyte traffic and their relevance to certain mechanisms of immuno-regulation has attracted much attention (De Lisi, 1977; Bell, 1978; Sprent, 1978). However, Hammond (1975) developed a mathematical model for circulatory lymphocytes in the spleen using the marginal zone, white pulp, and red pulp as compartments. Experimental results concerning lymphocyte traffic and a discussion of its immunological significance is given by Ford (1975) and Rannie and Ford (1977), among others. Interesting overviews of lymphocyte circulation are given by DeSousa (1977) and by McConnel, (1975). While lymphocytes are circulated throughout virtually every organ of the body, relatively few organs are significant from the immunological point of view, except for special problems.

A compartmentation of the immune system according to the most relevant organs includes bone marrow, blood, spleen thymus, lymph and lymph nodes, and gut-associated lymphoid tissue (GALT). The GALT includes the tonsils, small intestinal Peyer's patches, appendix, and peritoneal cavity. Bone marrow is the source of multipotential stem cells or precursor cells for the immune process. Spleen and lymph nodes are important locations for antibody–antigen reactions. Blood and lymph are important transport media, but also represent significant storage media for cells and molecules. Stem cells migrate from bone marrow to thymus and spleen, back via blood to GALT and lymph, and back to blood again.

In order to keep the model as tractable as possible and because of a lack of consistent experimental data, not all details of the migration patterns are taken into account. In the following discussion the basic assumptions are:

- The GALT compartment is neglected since one of the most significant roles of GALT is the generation of a particular class of antibodies (IgA) which is not considered here.
- Plasma cells (fully differentiated lymphocytes) and antigens do not recirculate in the blood compartment.
- T cells and macrophages are assumed to be present in sufficient quantities to induce a normal immune response, but their dynamics are neglected, and, consequently the thymus is neglected.
- The product of antibody–antigen reaction, that is, the immune complex, is assumed to be removed shortly after it is formed. Consequently, the immune complex density is not considered.

- As above, a distinction according to different classes of antibodies is not made here.

The state equations for the multicompartmental model are presented in this section, assuming that during the migration of lymphocytes and molecules the process dynamics throughout the compartments, to varying degrees, may be approximated in a similar manner to the single-compartment B model derived above. Here, an explanation of notation is in order. In the equations, the first subscript of the variables, the numbers 1, 2, 3, and 4, refers to immunocompetent cells, plasma cells, antibodies, and antigens, respectively. The second subscript, the letters b, s, l, and o, stands for blood, spleen, lymph and lymph nodes, and external compartments, respectively. Then the immune system is approximated by appropriate initial states and

$$\frac{dx_{1b}}{dt} = a_{1bs}x_{1s} + a_{1b}x_{1b} - (a_{1sb} + a_{1lb} + a_{10b})x_{1b} + v_1$$

$$\frac{dx_{1s}}{dt} = a_{1sb}x_{1b} - (a_{1bs} + a_{10s})x_{1s} + \alpha p_{ss}(1 - 2p_{ds})x_{1s}$$

$$\frac{dx_{1l}}{dt} = a_{1lb} - (a_{1bl} + a_{10l})x_{ls} + \alpha p_{sl}(1 - 2p_{dl})x_{1l}$$

$$\frac{dx_{2s}}{dt} = 2\alpha p_{ss}p_{ds}x_{1s} - a_{20s}x_{2s}$$

$$\frac{dx_{2l}}{dt} = 2\alpha p_{sl}p_{dl}x_{1l} - a_{20l}x_{2l}$$

$$\frac{dx_{3b}}{dt} = a_{3bl}x_{3l} - (a_{3lb} + a_{30b})x_{3b}$$

$$\frac{dx_{3s}}{dt} = \alpha x_{2s} - kcx_{4s}x_{3s} - (a_{3ls} + a_{30s})x_{3s}$$

$$\frac{dx_{3l}}{dt} = \alpha x_{2l} - kcx_{4l}x_{3l} + a_{3lb}x_{3b} + a_{3ls}x_{3s} - (a_{3bl} + a_{30l})x_{3l}$$

$$\frac{dx_{4s}}{dt} = a_{40s}x_{4s} - kc_k x_{3s}x_{4s}$$

$$\frac{dx_{4l}}{dt} = a_{40l}x_{4l} - kc_k x_{3l}x_{4l} \tag{5.1}$$

where a_{kji} denotes the transfer rate coefficient of material k to the jth compartment from the ith compartment. The terms p_{ss} and p_{sl} are the probabilities that antigen stimulates a cell in the spleen and lymph node, respectively: p_{ds} and p_{dl} are the probabilities that an ICC differentiates into a plasma cell in the spleen and a lymph node, respectively.

These probability terms are approximated by equations (3.4) and (3.5) in Chapter 3, for the spleen and lymph. Other parameters in the model are defined in Chapter 3. Antigen stimulation is introduced by non-zero initial conditions.

A simulation of the compartmental model of equation (5.1) for a mouse injected with a common experimental antigen, sheep's red blood cells (SRBC), is presented in *Figure 5.2*. The figure shows a comparison of the model simulation with the experimental data for a mouse with and without a spleen. The major differences between the simulation and the experimental antibody level seem to come from a switchover in antibody class which the model neglects.

A sensitivity analysis of the models and the simulation shows the following:

- Removal of the spleen results in an attenuation of antibody production by about a factor of three.
- The k term is a very critical parameter. At the same time, while c cannot be dropped in most cases, it need not be estimated accurately.
- The term α is somewhat critical, the other parameters are of less significance in defining model dynamics.

Following in the path of Mohler (1974) and Smith and Mohler (1976), conditions of compartmental accessibility may be derived which are necessary for a minimum model realization from input–output tracer data. It can be readily seen from standard linear system controllability and observability conditions for tracer dynamics that the blood compartment is excellent for tracer insertion and observation. The spleen or the lymph, however, are not recommended as single compartments of accessibility as a consequence of the relatively minor direct migration from the spleen to lymph.

It is obvious that future models should consider switchover. Also, memory cells have been shown to recirculate throughout the lymphoid system, playing a key role in immunity and enhanced secondary response. Time delays in certain organs (such as for B cells in the spleen) are significant. Conceptually, it is not difficult to include these complications with the compartmental model shown above. However, there is a lack of consistent data, which are necessary to determine key parameters for such a model at this time. Special problem analyses may very well require subcompartmentation of the spleen and lymph nodes.

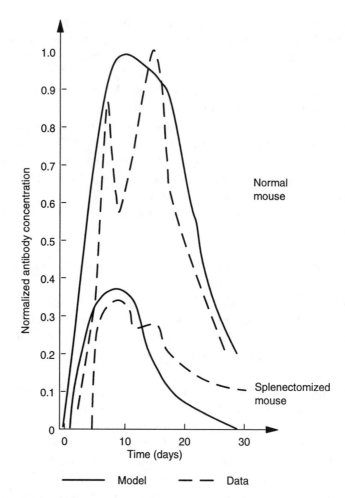

Figure 5.2. Normal and splenectomized antibody response of a mouse to sheep's red blood cells.

For example, certain tumors develop at specific nodes and T–B–M$_\phi$ interactions occurring for only a short period of time take place in the white pulp compartment of the spleen. Also, the lungs receive a high concentration of lymphocytes early in the response and are important in the early removal of antigen. An intensified program of collaboration between analysts and experimenters is necessary to understand these processes better.

5.2 Recirculating Lymphocytes

Basically, there are two types of lymphocyte migration:

- Homing, in which the cells move from one site to another.
- Recirculation, in which there is continuous movement between, and through, the lymphoid tissue.

The dynamics of the latter are studied here, following the pattern of the study of Farooqi and Mohler (1989). The anatomy and physiology of recirculation is studied by Govans and Knight (1964), Ford (1969), Hall *et al.*, (1976), and DeSousa (1977).

Factors that influence cell entry into the recirculating lymphocyte pool (RLP) are not fully understood. It is generally thought that cells leaving the primary lymphoid tissue undergo processing in the secondary lymphoid tissue (e.g., contact with an antigen) before entering the RLP (Sprent, 1977). This assumes that the virgin cells are short-lived and that they die soon or that they come into contact with an antigen, thereby eliciting a primary immune response to the antigen. This encounter determines the triggering of some of the responding cells to form long-lived, recirculating memory cells. Thus, perhaps all or most of the recirculating lymphocytes are memory cells, and any response they participate in is a secondary response.

A very brief review of recirculation follows. For more of the relevant details see Ford (1975, 1969), Ford and Govans (1969), DeSousa (1977, 1981), and Sprent (1973, 1977). Normally, there are very few lymphocytes in non-lymphoid tissue, so that recirculation pertains more to lymphoid tissue, although conditions can be created under which traffic 20–30 times above the baseline can occur in the former. It is well known that the lymphocyte output of the thoracic duct is mostly T cells (65–80% in rats) and that they recirculate faster than B cells (Ford, 1975). T cells recirculate through the body in about 18 hours on average. Of these, they spend only about one hour in blood or lymph (Parrott and Wilkinson, 1981). Various antigenic and non-antigenic factors can influence the migration of lymphocytes.

Antigenic stimulation causes increased vascularization in the appropriate area, increased vascular permeability, increased blood flow, and, consequently, increased lymphocyte traffic. There is also directly or indirectly induced antigen-oriented locomotion. This may be caused by lymphokines. Other factors which influence migration include: cell-mediated immunity, inflammation, and temperature. The most important factors that influence lymphocyte migration are:

- Hemodynamic.
- Physico-chemical interaction of cells and vascular endothelium.

A cell adhering to the endothelium is subjected to shear due to blood flow and also to an effective frictional drag. Thus, acceleration and cell concentrations are functions of distance along the vessel axis, so there is a concentration gradient. Then, by Fick's first law of diffusion, a flow exists which is proportional to the concentration differences across the interface.

5.3 Experimental Base for the Model

Smith and Ford (1983) studied the recirculating lymphocytes under conditions as close as possible to the physiological conditions. AO rats, adult male donors, and adult female recipients were used. In vitro labeling of thoracic duct lymphocytes was carried out with sodium-[^{51}Cr]-chromate, then they were passed from blood to lymph in an intermediate rat to ensure that only live recirculating lymphocytes were used. Five rats were sacrificed at most of the thirteen sample time periods. After removal of the 13 relevant organs (namely, blood, lungs, spleen, liver, right and left popliteal LN, coeliac LN, superficial cervical LN, deep cervical LN, mesenteric LN, Peyer's patches, gut, and bone marrow), the results of scintillation counting were calculated as a percentage of injected dose per organ and per gram of tissue. The published results are the means and standard deviations of activity expressed as a percentage of the injected dose.

One minute after injection, the lungs and blood each had approximately 40% of the injected dose, the liver had 13%, and there was very little elsewhere. After two minutes, the course of radioactivity in the lungs followed that in the blood. Up to about 30 minutes it fell sharply in both and substantially in the liver. In the spleen and in the LNs and Peyer's patches the concentration of the label steadily increased. Lymphocyte localization in the gut and bone marrow seems flat from 30–60 minutes on. The network of interconnections for the organs of the RLP is shown in *Figure 5.3*. Of the superficial and deep cervical LNs (SCLNs, a total of about 550 mg), a portion (350 mg) are associated with efferent lymphatics and the rest (200 mg) drain other tissues.

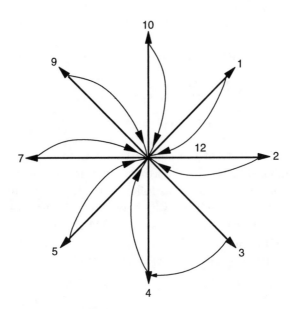

Figure 5.3. Connectivity diagram for the nine-compartment model. (For an explanation of the numbers, see *Figure 5.1*.)

5.4 Model Development

Several models were developed and their response compared with the data from Smith and Ford (1983). The model presented here was developed by Farooqi and Mohler (1989).

5.4.1 A linear, time-invariant model

The model is based on the connectivity diagram given in *Figure 5.1* with 12 compartments. The afflux of radioactivity from each compartment is assumed to be proportional to the amount of radioactivity (labeled lymphocytes) present in the compartment. A mass-balance analysis with intercompartmental flow rates approximated by a linear relation of concentration differences (Fick's first law of diffusion) yields the following model:

$$\frac{dx_i(t)}{dt} = \alpha_i x_{12}(t) - \beta_i g_i(t) x_i(t), \quad i = 1, \ldots, 5, 7, 9, 10$$

$$\frac{dx_6(t)}{dt} = \alpha_6 x_{12}(t) - \beta_6 g_6(t) x_6(t) + \beta_7 g_7(t) x_7(t)$$

$$\frac{dx_8(t)}{dt} = \alpha_8 x_{12}(t) - \beta_8 g_8(t) + \beta_9 g_9(t) x_9(t) + \beta_{10} g_{10}(t) x_{10}(t)$$

$$\frac{dx_{11}(t)}{dt} = \beta_4 g_4(t) x_4(t) - \beta_{11} g_{11}(t) x_{11}(t)$$

$$\frac{dx_{12}(t)}{dt} = -\sum_{i=1}^{11} \frac{dx_i(t)}{dt}$$

$$g_i(t) = 1, \quad i = 1, \ldots, 11 \tag{5.2}$$

where $x_i(t)$ is the state of the compartment i, that is, the percentage of activity (labeled cells) in the ith compartment relative to the injected dose. The subscripts are as follows: 1 – lungs; 2 – bone marrow; 3 – spleen; 4 – liver; 5 – SCLN with efferent lymphatics; 6 – SCLN with other tissues; 7 – miscellaneous tissues; 8 – MLN; 9 – gut; 10 – Peyer's patches; 11 – CLN; and 12 – blood. The terms α_i and β_i are constant parameters. Writing equation (5.2) in vector form

$$\frac{dX(t)}{dt} = AX(t) \tag{5.3}$$

where $X(t) = [x_1(t), \ldots, x_{12}(t)]^T$. The initial values, $X(0)$, are $[0, 0, 0, 0, 0, 0, 0, 0, 0, 0, 0, 100]^T$ and

$$A = \begin{bmatrix}
-\beta_1 & & & & & & & & & & & \alpha_1 \\
& -\beta_2 & & & & & & & & & & \alpha_2 \\
& & -\beta_3 & & & & & & & & & \alpha_3 \\
& & & -\beta_4 & & & & & & & & \alpha_4 \\
& & & & -\beta_5 & & & & & & & \alpha_5 \\
& & & & & -\beta_6 & \beta_7 & & & & & \alpha_6 \\
& & & & & & -\beta_7 & & & & & \alpha_7 \\
& & & & & & & -\beta_8 & \beta_9 & \beta_{10} & & \alpha_8 \\
& & & & & & & & -\beta_9 & & & \alpha_9 \\
& & & & & & & & & -\beta_{10} & & \alpha_{10} \\
& & & \beta_4 & & & & & & & -\beta_{11} & 0 \\
\beta_1 & \beta_2 & \beta_3 & 0 & \beta_5 & \beta_6 & 0 & \beta_8 & 0 & 0 & \beta_{11} & -\alpha_{12}
\end{bmatrix} \tag{5.4}$$

where $-\alpha_{12} = \sum_{i=1}^{10} \alpha_i$. As seen from the connectivity diagram (*Figure 5.1*), matrix A is strongly connected and is thus irreducible (Lancaster, 1969). The column sums are zero, thus making it column-diagonally dominant. The diagonal elements are negative and all the off-diagonal elements are non-negative. Thus, A is a closed compartmental system (Anderson, 1983) and essentially non-negative (Birkhoff and Varga, 1958): that is, the solution, $x(t)$, of the matrix in (5.4) is non-negative for $t \geq 0$ because A is irreducible and essentially non-negative. Consequently, negative diagonal elements ensure boundedness of the solution $x(t)$. The real parts of the eigenvalues of A are non-positive, but because A is closed compartmentally its spectral radius is unity (Jacquez, 1972). It can be shown that one eigenvalue is indeed zero and that no purely imaginary eigenvalues exist. Also, it was shown that autonomous, closed chemical reaction systems which may be represented by the matrix in (5.4) have a unique, stable equilibrium point at which the forward and reverse transfer rates between any two compartments are exactly equal (Shear, 1968).

Since all the compartments are input accessible and output measurable, the model [given by equation (5.3)] is identifiable. This can also be shown from the results of Smith and Mohler (1976), where it is seen that labeled cells injected into the thoracic duct and labeled counts made there are sufficient (theoretically) for minimum-order model realization. Matrix A (i.e., the parameters α_i and β_i) are estimated using physiologically reasonable guesses initially and then optimizing the parameters with Powell's technique, using a least-squares criterion with reference to the experimental data (Powell, 1964). The parameters obtained are as given in *Table 5.1*.

A linear, time-invariant model and a linear, constant-coefficient time-delay are also derived by Farooqi and Mohler (1989). Since the real-world system is non-linear in nature, however, it is logical to expect a non-linear model to fit the data better. Such a model follows, but in this case the network of compartments was simplified by combining all the lymph nodes together. The homogeneity of the structure and function of LNs is assumed in order to do this. This simplifies the migration connectivity diagram to that shown in *Figure 5.3*. This is a closed, strongly connected system. The model is given by:

$$\frac{dx_i}{dt} = -\beta_i x_i + \alpha_i x_{12}, \quad i = 1, \ldots, 7$$

$$\frac{dx_2}{dt} = a_{22}(t)x_2 + \alpha_2(1 - \gamma_2 x_2)x_{12}$$

$$\frac{dx_3}{dt} = a_{33}x_3 + \alpha_3 x_{12}$$

$$\frac{dx_4}{dt} = -a_{33}(t)x_3 - \beta_4 x_4 + \alpha_4(1 - \delta_4 x_4)x_{12}$$

$$\frac{dx_i}{dt} = -a_{ii}(t)x_i + [\alpha_i + \gamma_i(1 - \delta_i x_i)x_i]x_{12}, \quad i = 5, 9, 10$$

Table 5.1. Estimated values of parameters α_i and β_i (Powell, 1964).

i	1	2	3	4	5
α_i	1.800	0.0200	0.0550	0.2500	0.0045
β_i	2.000	0.0160	0.0300	0.0700	0.0002

i	6	7	8	9	10	11
α_i	0.0020	0.0150	0.0030	0.0150	0.0040	0
β_i	0.0300	0.0040	0.0010	0.0050	0.0060	1.000

Table 5.2. Estimated parameter values for the model given by equation (5.5).

i	1	2	3	4
α_i	1.500	0.0040	0.0550	0.5000
β_i	2.000	0.0160	0.0015	0.4200
γ_i	–	0.5500	0.0075	–
δ_i	–	–	–	0.0550
$\tau_{i,\min}$	–	360.0	150.0	–

i	5	7	9	10
α_i	0.0090	0.0250	0.0015	0.0011
β_i	0.0010	0.0020	0.0030	0.0006
γ_i	0.0035	–	0.0100	0.0080
δ_i	0.0470	–	1.000	0.1600
$\tau_{i,\min}$	720.0	–	300.0	540.0

$$
\begin{aligned}
\frac{dx_{12}}{dt} &= \sum_{i=1,4,7} \beta_i x_i - \sum_{i=2,5,9,10} a_{ii}(t)x_i - \alpha_{12}x_{12} \\
\alpha_{12} &= \alpha_1 + \alpha_2(1 - \gamma_2 x_2) + \alpha_3 + \alpha_4(1 - \delta_4 x_4) \\
&\quad + \alpha_5 + \gamma_5(1 - \delta_5 x_5) + \alpha_7 + \alpha_9 + \gamma_9(1 - \delta_9 x_9) \\
&\quad + \alpha_{10} + \gamma_{10}(1 - \delta_{10}x_{10})x_{10} \ .
\end{aligned} \tag{5.5}
$$

The time-variant parameters are given by

$$
\begin{aligned}
a_{22}(t) &= -\beta_2(1 - e^{-t/\tau_2}) \\
a_{33}(t) &= -[\beta_3 + \gamma_3(1 - e^{-t/\tau_3})] \\
a_{ii}(t) &= \beta_i(1 - e^{-t/\tau_i}), \quad i = 5,9,10 \ .
\end{aligned} \tag{5.6}
$$

The parameters α_i, β_i, and γ_i in the above models represent directional permeabilities that are proportional to flow rates in the various circulatory vessels and organs in different regions of the body. The term τ_i represents capacitive delays.

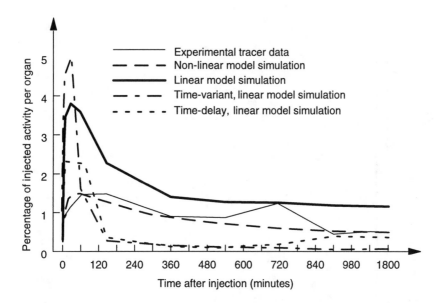

Figure 5.4. Lymphocyte response of bone marrow in rats.

Here, the state product terms x_1 and x_2 are approximations of the non-linear terms – predominantly diffusion-rate saturation such as $\alpha_i x_{12}/(1 + \gamma_i x_i)$. Of course, equation (5.5) may be written in the state form: $X(t) = [x_1(t), x_2(t), x_3(t), x_4(t), x_5(t), x_7(t), x_9(t), x_{10}(t), x_{12}(t)]^T$.

The parameters were estimated using physiologically reasonable guesses initially and varying these within a small range by trial and error until as small a difference as possible between the estimates and actual data was obtained. Estimation using those techniques employed for the previous models yielded results very close to the ones obtained by trial and error. The estimated parameter values are given in *Table 5.2*.

5.5 Simulation

The models mimic the data with various degrees of success. In *Figures 5.4* and *5.5* their outputs are compared with experimental data for the bone marrow and lungs of rats, respectively. Similar comparisons were made for the other compartments but are not presented for the sake of brevity.

In general, the fit of the time-delay model is better than the other two linear models, but it is not as good an approximation as the more

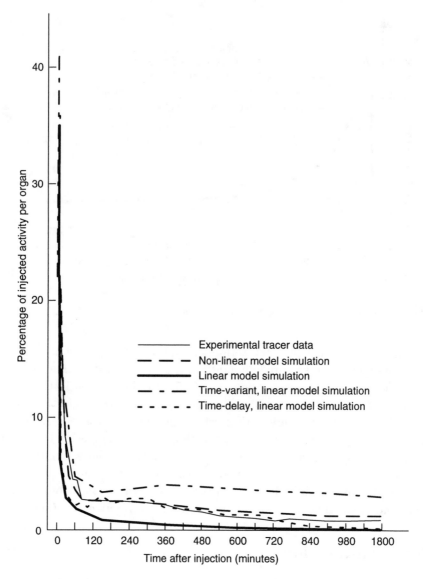

Figure 5.5. Lymphocyte response in the lungs of rats.

complicated non-linear model. The isotope used by Smith and Ford to radiolabel the lymphocytes preferentially tags T cells. B cells and T cells do not experience recirculation in the same way. Their proportions are different in different organs; they follow different routes within an organ; they have a different tempo of recirculation (B cells are slower); and they have different sojourn times within organs (e.g., larger for B cells in LNs). Thus, the models presented here represent the broad category of lymphocytes and do not separate B and T cells. All the parameters (α_i, β_i, γ_i, τ_i, etc.) refer to the broad category and not to one of the subcategories. In fact, data for B and T cells are not available separately. Modeling B and T cells separately would give a better fit to the real system. Consequently, certain estimated values may not be very realistic.

After a rapid lymphocyte exchange in the lungs during the first couple of minutes after injection, the level of lymphocytes decreases for the next few hours. As lymphocytes return to the blood, the decay ceases between 1 and 2.5 hours after injection and is followed by a slow rise to near equilibrium after 6 hours. Approximately 40% of the injected lymphocytes are found in the spleen after about 30 minutes. This is followed by a decay phase. Modeling the liver is particularly complicated because there might be three independent phenomena involved in lymphocyte migration in that organ. First, there is intravascular pooling, similar to the situation in the lungs, which results in a rapid initial response. Then there is genuine recirculation from the blood to the liver, to the coeliac lymph nodes, to the thoracic duct, and back to the blood again. Finally, there is an accumulation of damaged and dying cells in the liver. This involves up to 2% of the total population daily. Still, this can be a substantial part of the liver response itself. From the measured data, it is apparent that the basic premise of cell-population conservation isn't valid within the organs studied. In other words, part of the count is lost due to non-recirculation and removal, natural death, and migration to organs not measured. In the model these unaccountable cells are lumped into one compartment: the "other tissue" compartment. Obviously, there is a great deal of uncertainty associated with this loss, as well as with the migration coefficients which may cause large data variances. The latter, for the major part, is due to randomness in the complex flow phenomena, as well as randomness in the physical properties of both the cells (T and B lymphocytes) and the subjects (rats) sampled. Consequently, a stochastic model for recirculating lymphocyte distribution was derived. Unfortunately, due to the cell conservation assumption, the circulatory model is not asymptotically stable, and hence the stochastic model is not secondary (statistically) stable. Therefore, difficulties in deriving optimal parameter estimates from limited sample data precluded model accuracy.

In spite of all the above-mentioned problems, the analysis which is presented here does successfully decouple the recirculating (passive) lymphocyte dynamics from the dynamic response (and active migration) which is due to antigenic (alien substance) stimulation.

Appendix A

Modeling Delay Using Ordinary Differential Equations

In Chapters 2 and 4 an ordinary differential equation (ODE) with delay was used to model a disease process. The terms with delay describe the division and differentiation of the lymphocytes. Such a cascade process of cloning, for example, plasma cells takes several hours to several days. Here, we discuss the way in which we can modify the model with delay by using ODE.

Let the term $u_i(t)$, $i = 1, 2, \ldots, k$, denote the number of cells in state i and rewrite the equation for $C(t)$ from Chapter 2 in the form

$$\frac{d}{dt} = \xi(m_t)u_k(t) - \mu_c(C(t) - C^*) , \tag{A.1}$$

where $u_k(t)$ is the number of cells which have passed through k stages: that is,

$$\begin{aligned}
\frac{d}{dt}u_1(t) &= \gamma[F(t)V(t) - u_1(t)], \quad u_1(0) = 0 \\
\frac{d}{dt}u_2(t) &= \gamma[u_1(t) - u_2(t)], \quad u_2(0) = 0 \\
&\quad\vdots \\
\frac{d}{dt}u_k(t) &= \gamma[u_{k-1}(t) - u_k(t)], \quad u_k(0) = 0 \\
\gamma &= \text{constant} > 0 .
\end{aligned} \tag{A.2}$$

From the mathematical point of view such modification is based on the
following.

Theorem A.1

Let $x(t)$ be a continuous function on the interval $[-\tau, T]$, $0 < \tau < T$,
which has $(k+1)$ derivatives. Then,

$$u_k(t) = x(t - \tau) = O(\tau^2/k) ,\tag{A.3}$$

where

$$\frac{\tau}{k}\frac{d}{dt}u_1(t) = x(t) - u_1(t), \quad u_1(0) = x(-\tau/k)$$

$$\frac{\tau}{k}\frac{d}{dt}u_2(t) = u_1(t) - u_2(t), \quad u_2(0) = x(-2\tau/k)$$

$$\vdots$$

$$\frac{\tau}{k}\frac{d}{dt}u_k(t) = u_{k-1}(t) - u_k(t), \quad u_k(0) = x(-\tau)\tag{A.4}$$

Proof (Zuev, 1988)

It is obvious for $t = 0$. Let $T > 0$, now we can write the solution of
the equations in (A.4) in the series form. For the first equation we
have

$$u_1(t) = \sum_{i=0}^{\infty}\varepsilon^i y_i(t), \quad \varepsilon = \tau/k, \quad k > \tau ,\tag{A.5}$$

where $y_i(t)$, $i = 0, 1, \ldots$, are unknown functions. Substituting (A.5)
in the first equation of (A.4), we obtain

$$y_0(t) = x(t)$$
$$y_1(t) = -x^{(1)}(t)$$
$$y_3(t) = -x^{(2)}(t) \ldots\tag{A.6}$$

where

$$x^{(n)}(t) = \frac{d^n}{dt^n}x(t) .\tag{A.7}$$

Therefore,

$$u_1(t) = \sum_{i=0}^{n}(-\varepsilon)^i x^{(i)}(t) + O(\varepsilon^{n+1})\tag{A.8}$$

(for details see Trenogin, 1980).

From the other side of the equation, for the term $x(t - \varepsilon)$ we can write the Taylor expansion

$$x(t - \varepsilon) = x(t) - \varepsilon x^1 + \frac{\varepsilon^2}{2} - x^2(t) + O(\varepsilon^2) . \tag{A.9}$$

Correlating equations (A.8) and (A.9), we find that

$$u_1(t) = x(t - \varepsilon) + O(\varepsilon^2) . \tag{A.10}$$

Analogously, for the second equation we have

$$\begin{aligned} u_2(t) &= u_1(t) - \varepsilon u_1^1(t) + \varepsilon^2 u_1^2(t) + \cdots \\ u_2(t) &= u_1(t) - \varepsilon u_1^1 + \frac{\varepsilon^2}{2} u_1^2(t) + \cdots \end{aligned} \tag{A.11}$$

or

$$u_2(t) = u_1(t - \varepsilon) + O(\varepsilon^2) . \tag{A.12}$$

In the general form,

$$u_i(t) = u_{i-1}(t - \varepsilon) + O(\varepsilon^2), \quad i = 1, 2, \ldots, k . \tag{A.13}$$

From equation (A.13) we have

$$u_k(t) = x(t - \tau) + O(\tau^2/k) . \tag{A.14}$$

Now, based on this theorem, it is possible to compare the solutions of the system with delay.

$$\begin{aligned} \frac{d}{dt}x(t) &= f(x(t), x(t - \tau)), \quad x(s) = 0, \quad s \in [-\tau, 0] , \\ x(0) &= x_0, \quad t \in [0, T] \end{aligned} \tag{A.15}$$

and without delay,

$$\begin{aligned} \frac{d}{dt}y(t) &= f(y(t), u_k(t)), \quad y(0) = x_0, \quad t \in [0, T] \\ \frac{\tau}{k}\frac{d}{dt}u_1(t) &= y(t) - u_1(t), \quad u_1(0) = 0 \\ \frac{\tau}{k}\frac{d}{dt}u_2(t) &= u_1(t) - u_2(t), \quad u_2(0) = 0 \\ &\vdots \\ \frac{\tau}{k}\frac{d}{dt}u_k(t) &= u_{k-1}(t) - u_k(t), \quad u_k(0) = 0 \end{aligned} \tag{A.16}$$

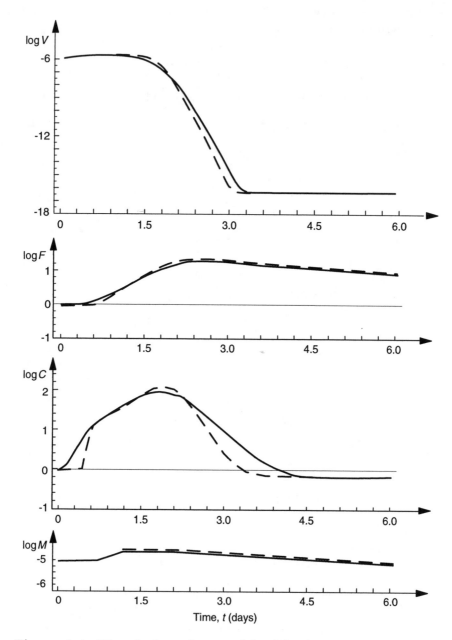

Figure A.1. The solutions of two models. (The dotted line corresponds to the model in Chapter 2; the solid line corresponds to the model in Appendix A.)

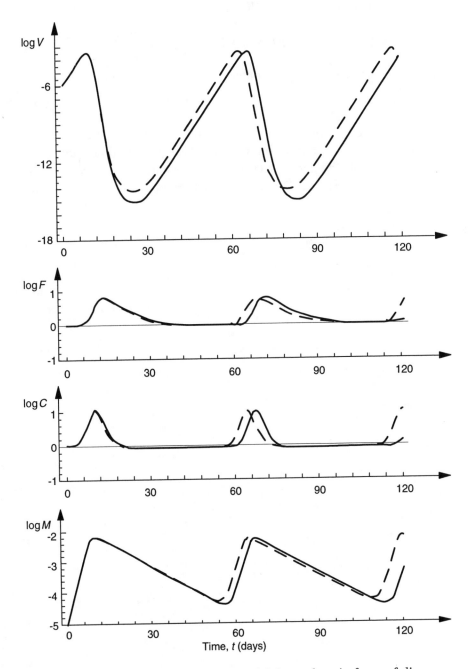

Figure A.2. The solution of the model for a chronic form of disease.

According to theorem A.1, for the system given in (A.16) we have

$$\frac{d}{dt} = y_k(t) = f(y(t), \ y(t-\tau) + \delta(t)), \quad \delta(t) = O(\tau^2/k) \ . \quad \text{(A.17)}$$

This allows us to estimate the difference

$$| \, x(t) - y(t) \, | \quad\quad\quad\quad\quad\quad\quad\quad\quad\quad\quad\quad\quad\quad \text{(A.18)}$$

using the theorems, for example, from Kamke (1969). As an illustration, the solutions of two models from Chapter 2 are presented in *Figures A.1* and *A.2*. The dotted lines correspond to the model from Chapter 2 and the solid line corresponds to the model considered here. In our example, $\tau = 1$ and $k = 2$.

References and Further Reading for Part I

Annual Reports, 1987–1991, Basel Institute for Immunology, Basel, Switzerland.

Anderson, D.H., 1983, Compartmental Modelling and Tracer Kinetics, *Lecture Notes in Biomathematics*, Vol. 50, Springer-Verlag, Berlin, Heidelberg, New York.

Anderson, R.M., and May, R.M., eds., 1991, *Infectious Diseases of Humans: Dynamics and Control*, Chapter 3, Oxford University Press, Oxford, UK.

Asachenkov, A.L., 1982, A Simple Model of Temperature-reaction Influence on the Immune-response Dynamics, pp. 41–47 in G.I. Marchuk, ed., *Matematicheskoe Modelirovanie v Immunologii Meditsine* (in Russian), Nauka, Novosibirsk, Russia.

Asachenkov, A.L., and Marchuk, G.I., 1979, Investigation of the Mathematical Model of a Disease and Some Results of Computer Simulation, pp. 69–78 in G.I. Marchuk, ed., *Proceedings of the IFIP Working Conference on Modelling and Optimization of Complex Systems*, Springer-Verlag, Berlin, Heidelberg, New York.

Asachenkov, A.L., and Mohler, R.R., 1989, A System Overview of Immunology, Disease and Related Data Processing, WP–89–31, International Institute for Applied Systems Analysis, Laxenburg, Austria.

Bell, G.I., 1970, Mathematical Model of Clonal Selection and Antibody Production I, *Journal of Theoretical Biology* **29**:191–232.

Bell, G.I., 1972, Mathematical Model of Clonal Selection and Antibody Production II, *Journal of Theoretical Biology* **33**:339–378.

Bell, G.I., 1973, Predator–Prey Equations Simulating an Immune Response, *Mathematical Biosciences* **16**:291–314.

Bell, G.I., 1978, Lymphocyte Traffic Patterns and Cell–Cell Interactions, pp. 341–375 in *Theoretical Immunology*, Marcel Dekker, New York, NY, USA.

Belykh, L.N., 1983, On the Computational Methods in Disease Models, pp. 79–84 in G.I. Marchuk and L.N. Belykh, eds., *Mathematical Modelling in Immunology and Medicine, Proceedings of the IFIP TC-7 Working Conference on Mathematical Modelling in Immunology and Medicine*, Moscow,

5–11 July 1982, North-Holland Publishing Company, Amsterdam, New York, Oxford.

Belykh, L.N., 1988, *Analiz Nekotorykh Matematicheskikh Modeley v Immunologii (Analysis of Some Mathematical Models in Immunology*: in Russian), p. 192, Nauka, Moscow, Russia.

Belykh, L.N., and Asachenkov, A.L., 1985, Mathematical Modelling of Infectious Diseases, pp. 12–79 in *Computational Processes and Systems*, Volume 3, Nauka, Moscow, Russia.

Belykh, L.N., and Marchuk, G.I., 1979, Chronic Forms of a Disease and Their Treatment According to Mathematical Immune Response Models, pp. 69–78 in G.I. Marchuk, ed., *IFIP Working Conference on Modelling and Optimization of Complex Systems*, Springer-Verlag, Berlin, Heidelberg, New York.

Benaceraff, B., and Unanue, E.R., 1979, *Textbook of Immunology*, University Park Press, Baltimore, MD, USA.

Birkhoff, G., and Varga, R.S., 1958, Reactor Criticality and Nonnegative Matrices, *Journal of SIAM* **6**(4):354–377.

Bocharov, G.A., 1983, Mathematical Model of Antiviral Immune Response of a Cellular Type, pp. 85–96 in G.I. Marchuk and L.N. Belykh, eds., *Mathematical Modelling in Immunology and Medicine*, North-Holland Publishing Company, Amsterdam, New York, Oxford.

De Boer, R.J., 1989, *Clonal Selection Versus Idiotypic Network Models of the Immune System: A Biomathematical Approach*, University of Utrecht, Utrecht, The Netherlands.

Bretscher, P.A., 1975, The Two-signal Model for B-cell Induction, *Transplant Review* **23**:37–48.

Bretscher, P.A., and Cohn, M., 1970, A Theory of Self-nonself Discrimination, *Science* **169**:1042–1049.

Brockett, R., 1975, On the Reachable Set for Bilinear Systems, pp. 54–63 in A.E. Ruberti and R.R. Mohler, eds., *Variable Structure Systems*, Springer-Verlag, Berlin, Heidelberg, New York.

Burnet, F.M., 1959, *The Clonal Selection Theory of Acquired Immunity*, Vanderbilt University Press, Nashville, TN, USA.

Chertkov, I.L., and Fridenshtein, A.Ya., 1977, *Kletochnye Osnovy Krovetvoreniya* (in Russian), Meditsina, Moscow, Russia.

Comincioli, V., Guerri, L., Serazzi, G., and Ugazio, A.G., 1979, A Mathematical Model of the Two-signal Theory for T-B Cell Cooperation, in C. Bruni, G. Doria, G. Koch, and R. Strom, eds., *Proceedings of the Working Conference on Systems Theory in Immunology*, 1978, Rome, Italy, Springer-Verlag, Berlin, Heidelberg, New York.

Druzchenko, V.E., 1982, Mathematical Model of the Immune Response of Humoral Type, in G.I. Marchuk and L.N. Belykh, eds., *Mathematical Modelling in Immunology and Medicine*, North-Holland Publishing Company, Amsterdam, New York, Oxford.

Farooqi, Z., and Mohler, R.R., 1989, Distribution Models of Recirculating Lymphocytes, *IEEE Transactions on Biomedical Engineering* **36**(3):355–362.

Fine, N.J., 1969, On the Walsh Functions, *Transactions of the American Mathematical Society* **65**:372–414.

Ford, W.L., 1969, The Kinetics of Lymphocyte Recirculation Within Rat Spleen, *Cell Tissue Kinetics* **2**:171–191.

Ford, W.L., 1975, Lymphocyte Migration and Immune Responses, *Progress in Allergy* **19**:1–59.

Ford, W.L., and Govans, J.L., 1969, The Traffic of Lymphocytes, *Seminars in Hematology* **6**:67.

Gillmore, R., 1974, *Lie Groups, Lie Algebras and Some of Their Applications*, J. Wiley, New York, NY, USA.

Govans, J.L., and Knight, E.J., 1964, The Route of Recirculation of Lymphocytes in the Rat, *Proceedings of the Royal Society B* **159**:257–282.

Greaves, M.F., Owen, J.J.T., and Raff, M.C., 1973, T and B Lymphocytes: Origins, Properties and Roles in Immune Responses, *Excerpta Medica*, Elsevier, New York, NY, USA.

Gruntenko, E.V., 1987, Immunitet i Vozniknovenie Zlokachestvennykh Opukholej (in Russian), Nauka, Novosibirsk, Russia.

Hall, J.G., Scollay, R., and Smith, M., 1976, Studies on Lymphocytes of Sheep. I. Recirculation of Lymphocytes Through Peripheral Lymph Nodes and Tissues, *European Journal of Immunology* **6**:117–120.

Hammond, B.J., 1975, A Compartmental Analysis of Circulatory Lymphocytes in the Spleen, *Cell Tissue Kinetics* **8**:153–169.

Hirschorn, R., 1973, Topological Semigroups, Sets of Generators and Controllability, *Duke Mathematical Journal* **40**:934–947.

Hoffman, G.W., 1979, A Mathematical Model of the States of Network Theory of Self-regulation, pp. 239–257 in C. Bruni, G. Doria, G. Koch, and R. Strom, eds., *Systems Theory in Immunology: Lecture Notes in Biomathematics*, Volume 32, Springer-Verlag, Berlin, Heidelberg, New York.

Jacquez, J.A., 1972, *Compartmental Analysis in Biology and Medicine: Kinetics of Distribution of Tracer-labeled Materials*, Elsevier Publishing Company, Amsterdam, The Netherlands.

Kamke, E., 1969, *Differentialgleichungen: Lösungsmethoden und Lösungen, Part I, Gewöhnliche Differentialgleichungen* (in German), Akademische Verlagsgesellschaft, Leipzig, Germany.

Karanam, V.R., Frick, P.A., and Mohler, R.R., 1978, Bilinear System Identification by Walsh Functions, *IEEE Transactions on Automatic Control* **23**:709–713.

Kaufman, M., 1988, Role Multistability in an Immune Response Model: A Combined Discrete and Continuous Approach, pp. 199–222 in A.S. Perelson, ed., *Theoretical Immunology, Part One, SFI Studies in the Science of Complexity*, Vol. II, Addison-Wesley, New York, NY, USA.

Kislyak, N.S., and R.V. Lenskaya, 1978, *Kletki Krovi u Detej v Norme i Patologii* (in Russian), Meditsina, Moscow, Russia.

Kolodziej, W.J., and Mohler, R.R., 1986, State Estimation and Control of Conditionally Linear Systems, *SIAM Journal on Control and Optimization* **24**:497–509.

Lancaster, P., 1969, *Theory of Matrices*, Academic Press, New York, NY, USA.

De Lisi, C., 1977, Some Mathematical Problems in the Initiation and Regulation of the Immune Response, *Mathematical Bioscience* **35**:1–26.

Lopukhin, Yu.M., and Petrov, R.V., 1974, New Classification of the Primary Immunologic Deficiency (in Russian), *Vestnic AMN SSSR*, No. 3, pp. 35–42.

Lopukhin, Yu.M., and Petrov, R.V., eds., 1975, *Vrozhdennye Immunodefitsitnye Sostoyaniya u Detej* (in Russian), Meditsina, Moscow, Russia.

McConnel, I., 1975, Structure and Function of Lymphoid Tissue, pp. 206–223 in M. Hobart and I. McConnel, eds., *A Course on the Molecular and Cellular Basis of Immunity*, Blackwell Scientific, Oxford, London.

Marchuk, G.I., ed., 1978, Some Mathematical Models in Immunology, pp. 41–62 in *Proceedings of the 8th IFIP Conference on Optimization Techniques*, 5–9 September, Würzburg, Germany, Springer-Verlag, Berlin, Heidelberg, New York.

Marchuk, G.I., 1979, Mathematical Immune Response Models and Their Interpretation, pp. 69–78 in *IFIP Working Conference on Modelling and Optimization of Complex Systems*, Springer-Verlag, Berlin, Heidelberg, New York.

Marchuk, G.I., 1983, *Mathematical Models in Immunology*, Optimization Software, Inc., Publications Division, New York, NY, USA.

Marchuk, G.I., 1992, *Mathematical Models in Immunology* (in Russian), Nauka, Moscow, Russia.

Marchuk, G.I., and Berbentsova, A.P., 1989, *Acute Pneumonia*, p. 302, Nauka, Moscow, Russia.

Marchuk, G.I., and Petrov, R.V., 1983, Mathematical Model of Antiviral Immune Response (in Russian), in G.I. Marchuk, ed., *Computational Processes and Systems*, Vol. 3., Nauka, Moscow, Russia.

Marchuk, G.I., and Zuev, S.M., 1987, Estimation of Response Model Parameters Based on Maximum Likelihood Method, pp. 91–98 in A. Germani, ed., *Stochastic Modelling and Filtering, Proceedings of the IFIP-WG 7.1 Working Conference*, 10–14 December 1984, Rome, Italy, Springer-Verlag, Berlin, Heidelberg, New York.

Marchuk, G.I., Asachenkov, L.A., Belykh, L.N., and Zuev, S.M., 1986, Mathematical Modelling of Infectious Diseases, pp. 64–81 in T. Hraba and G. Hoffman, eds., *Immunology and Epidemiology, Proceedings of International Conference*, 18–25 February 1985, Mogilani, Poland, Springer-Verlag, Berlin, Heidelberg, New York.

Marchuk, G.I., Belykh, L.N., and Zuev, S.M., 1987, Mathematical Models in Immunology, pp. 462–475 in R.E. Kalman, G.I. Marchuk, A.E. Ruberti, and A.J.Viterbi, eds., *Recent Advances in Communication and Control*

Theory, Optimization Software, Inc., Publications Division, New York, NY, USA.

Marchuk, G.I., Petrov, R.V., Romanyukha, A.A., and Bocharov, G.A., 1991, Mathematical Model of Antiviral Immune Response I: Data Analysis, Generalized Picture Construction and Parameter Evaluation for Hepatitis B, *Journal of Theoretical Biology* **151**:41–70.

Mohler, R.R., 1973, *Bilinear Control Processes*, Academic Press, New York, NY, USA.

Mohler, R.R., 1974, Biological Modelling with Variable Compartmental Structure, *IEEE Transactions on Automatic Control* **19**:922–926.

Mohler, R.R., 1991, *Nonlinear Systems, Vol. II: Application to Bilinear Control*, Prentice Hall, Englewood Cliffs, NJ, USA.

Mohler, R.R., Barton, C.F., and Hsu, C.S., 1978, T and B Cells in the Immune System, pp. 415–436 in *Theoretical Immunology*, Marcel Dekker, New York, NY, USA.

Mohler, R.R., Bruni, C. and Gandolfi, A., 1980, A Systems Approach to Immunology, *Proceedings of the IEEE* **68**:964–990.

Nisevich, N.I., Marchuk, G.I., Zubikova, I.I., and Pogozhev, I.B., 1984, *Mathematical Modelling of Viral Disease*, Optimization Software Inc., New York, NY, USA.

Nossal, G.J.V., 1969, *Antibodies and Immunity*, Basic Books, New York, NY, USA.

Nossal, G.J.V., and Schrader, J.W., 1975, B Lymphocyte–Antigen Interactions in the Initiation of Tolerance or Immunity, *Transplant Review* **23**:136.

Parrott, D.M.V., and Wilkinson, P.C., 1981, Lymphocyte Locomotion and Migration, *Progress in Allergy* **28**:193–284.

Perelson, A., Mirmirani, M., and Oster, G.F., 1976, Optimal Strategies in Immunology: B-Cell Differentiation and Proliferation, *Journal of Mathematical Biology* **3**:325–367.

Petrov, R.V., 1976, *Immunologiya i Immunogenetika* (in Russian), Meditsina, Moscow, Russia.

Petrov, R.V., 1983, *Immunology* (in Russian), Meditsina, Moscow, Russia.

Petrov, R.V., Dozmorov, I.M., and Luzenko, C.B., 1985, Two-phase Character of Interaction of T Lymphocyte With Stem Cells (in Russian), *Immunology* **4**:16–19.

Powell, M.J.D., 1964, An Efficient Method of Finding the Minimum of a Function of Several Variables Without Calculating Derivatives, *The Computer Journal* **7**:303–307.

Rannie, G.H., and Ford, W.L., 1977, Physiology of Lymphocyte Recirculation in Animal Models, pp. 165–171 in *Proceedings of Excerpta Medica International Congress Series 423, Transplantation and Clinical Immunology*, Elsevier, New York, NY, USA.

Richter, P.H., 1978, The Network Idea and the Immune Response, pp. 539–569 in G.I. Bell, A.S. Perelson, and G.H. Pimbley, eds., *Theoretical Immunology*, Marcel Dekker, New York, NY, USA.

Roitt, I., 1974, *Essential Immunology*, Blackwell Scientific, Oxford, London.

Sapin, M.R., Yurina, N.A., and Etingen, L.E., 1978, *Lymph Node* (in Russsian), Meditsina, Moscow, Russia.

Schimpl, A., and Wecker, E., 1975, A Third Signal in B-Cell Activation Given by TRF, *Transplant Review* 23:176–188.

Shear, D.B., 1968, Stability and Uniqueness of the Equilibrium Point in Chemical Reaction Systems, *Journal of Chemical Physics* 48(9):4144–4147.

Skalko, Yu.I., 1983, *The Onset and Development of a Disease With Decreasing Efficiency of the Functioning of the Immune System* (in Russian), Preprint No. 69, Department of Numerical Mathematics of the USSR Academy of Sciences, Moscow, Russia.

Skalko, Yu.I., and Stepanenko, R.N., 1985, Modelling of Stimulator of Antibody Production Action (in Russian), in G.I. Marchuk, ed., *Computational Processes and Systems*, Nauka, Moscow, Russia.

Smith, M.E., and Ford, W.L., 1983, The Recirculating Lymphocyte Pool of the Rat: A Systematic Description of the Migratory Behavior of Circulating Lymphocytes, *Immunology* 49:83–94.

Smith, W.D., and Mohler, R.R., 1976, Necessary and Sufficient Conditions in the Tracer Determination of System Order, *Journal of Theoretical Biology* 57:1–21.

DeSousa, M., 1977, Cell Traffic, in P. Cuatrecasas and M.F. Greaves, eds., *Receptors and Recognition*, Series A, Volume 2, Chapman and Hall, London, UK.

DeSousa, M., 1981, *Lymphocyte Circulation: Experimental and Clinical Aspects*, John Wiley and Sons, New York, NY, USA.

Sprent, J., 1977, Recirculating Lymphocytes, pp. 43–111 in J.J. Marchalonis, ed., *The Lymphocyte: Structure and Functions*, Marcel Dekker, New York, NY, USA.

Sprent, J., 1978, Circulating T and B Lymphocytes of the Mouse: I, Migratory Properties, *Cell Immunology* 7:10–39.

Stepanenko, R.N., and Skalko, Yu.I., 1983, On the Mechanisms of Stimulator of Antibody Production Action, pp. 225–236 in G.I. Marchuk and L.N. Belykh, eds., *Mathematical Modelling in Immunology and Medicine, Proceedings of the IFIP TC-7 Working Conference on Mathematical Modelling in Immunology and Medicine*, 5–11 July 1982, Moscow, Russia, North-Holland Publishing Company, Amsterdam, New York, Oxford.

Takeuchi, Y., Aduchi, N., and Tocumaru, H., 1978, The Stability of Generalized Volterra Equations, *Journal of Mathematical Analysis and Application* 62:453–473.

Trenogin, V.A., 1980, *Functional Analysis* (in Russian), Nauka, Moscow, Russia.

Zinkernagel, R.M., 1976, H-2 Compatibility Requirement for Virus Specific T-Cell-Mediated Cytolysis, *Journal of Experimental Medicine* 143:443–473.

Zuev, S.M., 1988, *Statistical Estimation of Parameters of Mathematical Models of Disease* (in Russian), Nauka, Moscow, Russia.

Part II

Clinical Data Processing

Introduction

Appropriate methods for estimating model parameters are studied here in Chapters 6, 7, and 8. Then one particular method is applied to influenza data to give a different method of treatment in Chapter 9. The section ends with Chapter 10 where the individuality of patients is considered.

The study of disease in order to find methods of treatment is an important medical problem. As a rule, such problems are reduced to the study of the influence of external effects (drug action) on the process in question.

Let $x = (x^1, x^2, ..., x^n)^T$ be a state vector of the model: the terms x^i, $i = 1, ..., n$, are the indicators that we can measure in the clinic (quantity of viruses, antibodies, leukocytes, etc.). For each patient we have a set of measurements:

$$X = \{x_t, t \in \Theta\}, \quad x_t = (x_t^1, x_t^2, \ldots, x_t^n), \quad \Theta = (t_1, t_2, \ldots, t_N) \ .$$

Here, Θ is a set of instants of time at which the state vector is measured. If we have a group of m patients, there is a set of trajectories

$$\begin{aligned} X_m &= \{X^i, i = 1, 2, \ldots, m\} \\ X^i &= \{\vec{x}_t^i, t \in \Theta^i\} \ . \end{aligned}$$

The mathematical problem is to construct a statistic $\sigma(X_m) \in \mathbb{R}^L$ such that one can use classical statistical criteria for verification of the hypotheses on the possible influence of the effect in question. Assume that the model consists of a system of ordinary differential equations:

$$\frac{d}{dt} x_t = f(x_t, \alpha)$$

$$t \in [0, T], \quad x_0 = c \ , \tag{II.1}$$

where $x_t \in \mathbb{R}^n$ is a vector of state variables and $\alpha \in \mathbb{R}^L$ is a vector of coefficients.

The right-hand side of the model contains a priori information about the process studied. Thus the parameters of the model have a biological interpretation. Therefore, one can use the model given in equation (II.1) for data analysis. The estimate of the parameter vector $\alpha_m = \alpha(X_m)$, defined by X_m, is a statistic $\sigma(X_m)$, which we have to construct such that

$$\sigma(X_m) = \alpha(X_m) \ .$$

The main question is to determine the model parameters, α, using the observed data X_m. It should be noted that knowledge of the estimate itself is not of great importance from a practical (medical) point of view since we have to know the statistical properties of the estimate. Indeed, the problem of the external effect is reduced to the comparison of estimates $\alpha(X_m)$ and $\alpha(Y_m)$. Here, for example, X_m are data that take into consideration the external influence, while Y_m values do not. The question is how to compare estimates $\alpha(X_m)$ and $\alpha(Y_m)$ to establish the action effect. To this end we have to know something about the accuracy of the calculated estimates. This is why the traditional approach, such as the least-square method with the criterion

$$\min_{\alpha} J(\alpha) = \sum_{t \in \Theta} \sum_{i=1}^{m} (x_t^i - x_t(\alpha))^T (x_t^i - x_t(\alpha)) \ ,$$

where $x_t(\alpha)$ is the solution of (II.1), gives us only slightly or no useful information from a practical point of view. Such an approach does not take into account the probability properties of estimates. However, we need to know them. Therefore, let us define the problem in terms of model-based, clinical-data processing.

As a rule, the model describes only the main characteristics of change in the state vector. However, there are various internal and external factors that influence the process dynamics. Therefore, the real trajectories, as well as the general regularity, have a stochastic character that is not described by the model. Thus, the set of trajectories X_m may be considered as a set of realizations, defined on Θ, of some stochastic process with values in $\mathbb{R}^n$:

$$\tilde{x}_t = \{\tilde{x}_t(\omega), \ \omega \in \Omega, \ t \in [0, T]\} \ . \tag{II.2}$$

Assume that the model in equation (II.1) describes the mathematical expectation for a group of patients. It means that there exists a vector α^* such that

$$x_t(\alpha^*) = E\tilde{x}_t = \int_\Omega \tilde{x}_t(\omega)dP(\omega) \ ,$$

where $x_t(\alpha)$ is the solution of (II.1) and $x_t(\omega)$ is a stochastic process as in (II.2). The problem is to estimate α^* by using X_m. If $E\tilde{x}_t$, $t \in \Theta$ is known, then methods for the solution of non-linear problems should be used to calculate α^*. It should be noted that estimating the values of $E\tilde{x}_t$ for every $t \in \Theta$ requires a lot of data: this is very expensive in immunology and medicine. Another approach is to describe the random character of the observations by a stochastic model. The latter is discussed in this part with reference to Marchuk (1981, 1983), Zuev (1988), and Pogozhev (1988).

Chapter 6

The Problem of Parameter Estimation

6.1 Problem Statement

Consider the following system of ordinary differential equations (ODEs):

$$\frac{d}{dt}x_t = f(x_t, \alpha)$$
$$x_0 = c, t \in [0, T], \quad x_t \in \mathbb{R}^n, \quad \alpha \in \mathbb{R}^L . \tag{6.1}$$

[In this part the system of ordinary differential equations without delay is considered because the equation with delay may be transformed into the form of equation (6.1). (See Appendix A.)] Let the vector function $f(x, \alpha)$ be linear in α:

$$f(x, \alpha) = F(x)\alpha ,$$

where $F(x)$ is an appropriately smooth $n \times L$ function. Let $x_t(\alpha)$ denote the solution of equation (6.1) for $t \in [0, T]$. Furthermore, suppose that the interval of observation $[0, T]$ corresponds to the transition time of the modeled process. For example, the instant $t = 0$ corresponds to the time of antigen injection into the body, and T denotes the time required for reaction of the system to this perturbation. In particular, we can choose T as the time necessary for antigen removal and re-establishment of the initial (healthy) state. The proposition that the model describes the transition process means that there exists no $t \in [0, T]$ and $\alpha \in D \subset \mathbb{R}^L$ such that

$$\frac{d}{dt}x_t = f(x_t, \alpha) = 0 .$$

Except for this, we suppose the solution $x_t(\alpha)$, $\alpha \in D$, to be asymptotically stable. Let us denote $x_t(\alpha, t_0, x_0)$ as the solution of equation (6.1) for $t \geq t_0 \geq 0$ and the initial condition x_0. Recall that the asymptotic stability condition means that there exists some values ρ and σ, $\sigma \leq \rho$, such that when $|x_0 - y_0| < \sigma$,

$$\lim_{t \to \infty} |x_t(\alpha, t_0, x_0) - x_t(\alpha, t_0, y_0)| = 0 \ .$$

Moreover, for a linear approximation, the difference

$$\delta x_t = x_t(\alpha, t_0, y_0) - x_t(\alpha, t_0, x_0)$$

satisfies the system of equations

$$\begin{aligned}
\frac{d}{dt}\delta x_t &= f_x(x_t(\alpha, t_0, x_0), \alpha)\delta x_t \\
\delta x_0 &= y_0 - x_0
\end{aligned}$$

for sufficiently small δx_0 so that $\delta x_t \to 0$ when $t \to \infty$.

Here, $f_x(x, \alpha)$ is the Jacobian of the right-hand side of equation (6.1). The data on the dynamics of state variables are available in the form of results of clinical observations and immunological experiments. This set is given by

$$X_m = \{x_t^i, t \in \Theta, i = 1, 2, ..., m, x_t^i \in \mathbb{R}^n\}$$

where m is the number of trajectories.

Obviously, the real trajectories have a random character, and due to this they can not be realized by the model (6.1) by means of any choice of constant vector $\alpha \in D$. As mentioned above, the random character of trajectories is not only due to measurement error. So when we describe the real trajectories, we need to take into account the factors that have an influence on the process being studied but that were excluded from the description when we constructed the deterministic model. These factors did not determine the development of the process which induces the deviation of the trajectory from the general regular pattern. So the differences between trajectories are not principal, and we will consider them as purely random.

It is clear that we are dealing with qualitatively homogeneous trajectories that we can consider as the result of repetitions of the experiment with one organism. As a result of this, the behavior of real trajectories follows the regular function described by the deterministic model in equation (6.1) at $\alpha = \alpha^*$. We want to find this regular function by determining the value of the vector α^*. To estimate this vector we need to construct the functional $J(x_t(\alpha), X_m)$ such that

$$\max_{\alpha} J(x_t(\alpha), X_m) = J(x_t(\alpha_m), X_m) \ . \tag{6.2}$$

When $m \to \infty, \alpha_m \to \alpha^*$, where α^* is an unknown true value. The notation '$m \to \infty$' means that the number of independent trajectories is increasing and has no limits. It should be noted that the real trajectories are stochastic in character and don't belong to the set of solutions of equation (6.1). To construct the functional (6.2) we need to construct the stochastic model for the description of the observed state dynamics.

First of all, to describe the main idea, we will concentrate our attention on the simple model ($n = L = 1$, where n is the number of equations and L is the number of parameters):

$$\frac{d}{dt}x_t = -\alpha x_t, \quad x_0 = 1, \quad t \geq 0, \quad \alpha > 0 \ .$$

This is used for clinical data analysis, in particular for the evaluation of therapy efficiency (Nisevich *et al.*, 1984).

6.2 Clinical Data Processing

The problem connected with clinical data is the following. One approach to processing the observable data using the model supposes that the state variables are measurable. Unfortunately, the majority of these variables are difficult or impossible to measure in the clinic. As an example, consider the simple model of a disease given in Chapter 2. There are four state variables in this model (V, C, F, and m). We can easily measure common levels of antibodies and plasma cells, but specific antibodies and plasma cells form an unknown part of these levels. The measurement of antigen is more difficult, and we do not know how to measure the relative characteristic of a damaged organ or the degree of seriousness of the disease.

At the same time, physicians measure other characteristics forming a clinical state vector y where $y = (y^1, \ldots, y^n)^T$. By using these data, specialists can evaluate the state of the organism, choose the method of treatment, and investigate the process of rehabilitation of the afflicted organ. How can we use the information from clinically measured variables? One way to solve this problem is considered here and in Appendix B.1.

Let $t_{max} > t_0$ be the instant when the relative characteristic of organ damage, $m(t)$ (see Chapter 2), reaches its maximum value. For a favorable outcome (recovery) we can neglect the antigen (viruses) in the organism. Then, for $m(t)$ we have the following equation:

$$\frac{dm}{d\tau} = -\mu m$$

$$\tau \geq t_{max}, \quad m(t_{max}) = m_{max} \ .$$

Let us denote t by $t = \tau - t_{max}$, then

$$\frac{dm}{dt} = -\mu m, \quad t \geq 0, \quad m(0) = m_{max} \ . \tag{6.3}$$

As a rule, the time $t = 0$ corresponds to the moment the patient enters the hospital. According to the model given by equation (6.3), the decrease in organism damage is exponential in time. Any deviation from this regularity suggests unfavorable disease dynamics. To use this result in clinical practice, $m(t)$ should be related to values actually observed. It is natural to suppose that such values are in fact estimates of the severity of a patient's state given by clinicians. Let us introduce the estimate of the gravity of the disease by using the term s, where $s = 0, 1, ..., r$. The following scale is traditionally used by clinicians for classification of an organism's state:

- Healthy individuals, $s = 0$.
- Patients with a light form of affliction, $s = 1$.
- The mild form, $s = 2$.
- The serious form, $s = 3$.
- The serious form with unpredictable outcome, $s = 4$.

To identify the state of an organism according to this classification (in our notation the term s), physicians analyze the values of vector y components, $y^1, ..., y^n$. This means that s is a function of this vector. To find this function we can construct its estimate, $\phi(y)$, as a scalar function of the vector y: this is the so-called generalized index of the severity of a patient's state or the gravity index (GI). Methods for the derivation of GI are presented in Appendix B.1.

Experience using GI in the clinic shows that GI is similar to the variable M in the model given by equation (6.3). Also, with the help of GI one can successfully solve various practical problems such as estimating the degree of seriousness of the disease or the state of the organism by studying the disease dynamics and comparing the effectiveness of different therapies. A general framework of mathematical analysis of medical information was formulated by Marchuk (1984) and Zuev (1988). Clinical experience with such analyses are discussed by Pogozhev (1988) and Nisevich *et al.* (1984).

6.3 Stochastic GI Model

Now the dynamics for the favorable course of disease are studied. It should be noted that this concept was first introduced for clinical purposes, rather than mathematical, by Zubikova. After studying the disease history of patients with viral hepatitis, Zubikova and his fellow workers concluded the following. During the favorable disease history, the dynamics of the normalized GI, given by

$$x_t = \frac{\phi_t}{\phi_0}$$

$$\phi_t = \phi(y_t^1, \ldots, y_t^n)$$

are exponential on average: that is,

$$\overline{x}_t = Ex_t = \exp(-\lambda t) \ ,$$

where $\lambda = 0.12 \, \text{day}^{-1}$ and $t = 0$ is the instant at which the patient enters the hospital. This result corresponds to the model in equation (6.3). They also proposed to use the parameter λ as an integral characteristic of the recovery process. Naturally, it was then decided to estimate λ in the following model:

$$\begin{aligned} \frac{d}{dt}\overline{x}_t &= -\lambda \overline{x}_t \\ \overline{x}_0 &= 1, \ t \geq 0 \end{aligned} \tag{6.4}$$

which describes the average GI trajectories. Some GI trajectories are presented in *Figure 6.1*.

Correlating this data and the model in equation (6.4) shows that the deviations of real trajectories from the average dynamics in equation (6.4) may be described by variations in the parameter λ. At this level of modeling, we consider these variations as random. Therefore, let us consider the following stochastic model for the description of GI trajectories:

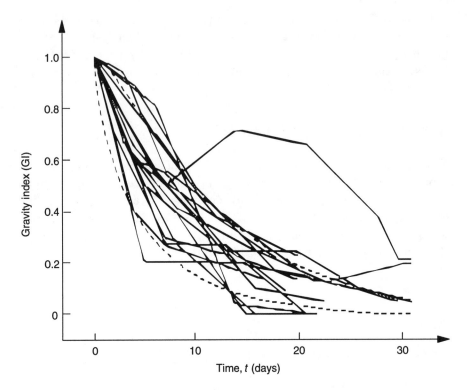

Figure 6.1. Typical dynamics of normalized GIs for viral hepatitis.

$$\frac{d}{dt}x_t^\varepsilon = -(\lambda + \zeta_t)x_t^\varepsilon$$
$$x_0^\varepsilon = 1, \ t \in [0,T]$$
$$\zeta_t = \{\zeta_t(\omega), \ t \in [0,T], \ \omega \in \Omega\}, \ E\zeta_t = 0, \ E\zeta_t^2 < \infty \ .$$

Here, the term ζ_t represents a random process which takes into account the individual deviations of each trajectory. The details of this representation are discussed in Chapter 7.

The recovery of the normal value, x_t, is a long process that takes time $T = 20$–30 days, but trajectory deviations from it have an interval of less than 1 day. In other words, considering that the random variable ζ_t is fast in comparison with the average dynamics of x_t, one can suppose

$$\frac{1}{T}\int_0^T \zeta_t dt \approx 0 \ . \tag{6.5}$$

This equality takes place in the limit as $T \to \infty$ when $\text{cov}(\zeta_t, \zeta_{t+\tau}) \to 0$ at $\tau \to \infty$. Introduce the process ξ_t such that $E\xi_t = 0, E\xi_t^2 < \infty$, $\text{cov}(\xi_t, \xi_{t+\tau}) \to 0$ as $\tau \to \infty$. Let $\zeta_t = \xi_{t/\varepsilon}$, where ε is a small parameter. Then condition (6.5) can be formulated in the following way. For any $T > 0$ and $\delta > 0$, the following limit relation takes place uniformly on t:

$$\lim_{\varepsilon \to 0} \mathbf{P} \left\{ \left| \frac{1}{T} \int_t^{t+T} \xi_{s/\varepsilon} ds \right| > \delta \right\} = 0 ,$$

where $P(A)$ is the probability of event A. So, we have the stochastic model for the description of experimental data:

$$\frac{d}{dt} x_t^\varepsilon = -(\lambda + \xi_{t/\varepsilon}) x_t^\varepsilon$$
$$x_0^\varepsilon = 1, \quad t \in [0, T] . \tag{6.6}$$

The term ε is of the order of the ratio of the characteristic times of random oscillations (hours) to recovery time (days, weeks). Now we can assume that in the model in equation (6.6) the value of ε is sufficiently small for us to use the properties of the limit process for x_t^ε at $\varepsilon \to 0$. Let us denote w_t^ε by

$$w_t^\varepsilon = \frac{1}{\sqrt{\varepsilon}} \int_0^t \xi_{s/\varepsilon} ds .$$

Then equation (6.6) takes the form

$$\frac{d}{dt} x_t^\varepsilon = -(\lambda + \sqrt{\varepsilon} \, \frac{d}{dt} w_t^\varepsilon) x_t^\varepsilon . \tag{6.7}$$

Theorem 6.1
Let the random process denoted by ξ_t have piecewise continuous (with probability 1) trajectories, zero mean, and satisfy the condition of strong mixing, with coefficient $\beta(\tau)$, such that

$$\int_0^\infty [\beta(\tau)]^{1/5} \tau \, d\tau < \infty .$$

Then, as $\varepsilon \to 0$, the process w_t^ε weakly converges on the interval $[0, T]$ to a Gaussian process w_t^0 with zero mean, independent increments, and covariance γt, where γ is the spectral intensity of the process ξ_t, that is

$$\gamma = \lim_{T \to \infty} \frac{1}{T} \int_0^T \int_0^T E\xi_t \xi_s \, dt \, ds .$$

This result is a particular case of a theorem proved by Ventcel and Freidlin (1975). Here, we explain it without strict proof. First, $E\xi_t = 0$ means that

$$Ew_t^\varepsilon = \frac{1}{\sqrt{\varepsilon}} E \int_0^t \xi_{s/\varepsilon} ds = \frac{1}{\sqrt{\varepsilon}} \int_0^t E\xi_{s/\varepsilon} ds = 0$$

and $Ew_t^0 = 0$. To find the covariance, rewrite the expression for $E(w_t^\varepsilon)^2$:

$$
\begin{aligned}
E(w_t^\varepsilon)^2 &= \frac{1}{\varepsilon} E \left(\int_0^t \xi_{s/\varepsilon} ds \right)^2 \\
&= \frac{1}{\varepsilon} E \int_0^t \int_0^t \xi_{s/\varepsilon} \xi_{\tau/\varepsilon} \, ds \, d\tau \\
&= \varepsilon \int_0^{t/\varepsilon} \int_0^{t/\varepsilon} E\xi_s \xi_\tau \, ds \, d\tau \ .
\end{aligned}
$$

Then, denoting T by $T = 1/\varepsilon$,

$$
\begin{aligned}
\lim_{\varepsilon \to 0} E(w_t^\varepsilon)^2 &= t \lim_{T \to \infty} \frac{1}{tT} \int_0^{tT} \int_0^{tT} E \, \xi_s \xi_\tau \, ds \, d\tau \\
&= t \lim_{\Theta \to \infty} \frac{1}{\Theta} \int_0^\Theta \int_0^\Theta E \, \xi_s \xi_\tau \, ds \, d\tau = \gamma t \ .
\end{aligned}
$$

By analogy, one can show that w_t^0 has independent increments, that is, for $t_0 < t_1 < t_2$,

$$E(w_{t_2}^0 - w_{t_1}^0)(w_{t_1}^0 - w_{t_0}^0) = 0 \ .$$

Therefore, the limit process w_t^0 has continuous trajectories, independent increments, zero mean, and covariance γt. Such a limit process is necessarily Gaussian. As long as the parameter ε is sufficiently small, then we can rewrite equation (6.7) as

$$
\begin{aligned}
\frac{d}{dt} x_t &= -\left(\lambda + B\frac{d}{dt} w_t \right) x_t \\
x_0 &= 1 \ ,
\end{aligned}
\tag{6.8}
$$

where $B = (\varepsilon\gamma)^{1/2}$ and w_t is a Wiener process.

6.4   Parameter Estimation

Let us consider the solution of parameter estimation for this stochastic model. First, represent equation (6.8) in the form

$$d\ln x_t = -\lambda dt + Bdw_t \ .\tag{6.9}$$

Then, for $s < t < \tau$,

$$\ln x_\tau - \ln x_t = -\lambda(\tau - t) + B(w_\tau - w_t)$$
$$\ln x_t - \ln x_s = -\lambda(t - s) + B(w_t - w_s)$$

and we obtain

$$\frac{(\ln x_\tau)/x_t + \lambda(\tau - t)}{(\tau - t)^{1/2}} = B\frac{w_\tau - w_t}{(\tau - t)^{1/2}}$$
$$\frac{(\ln x_t)/x_s + \lambda(t - s)}{(t - s)^{1/2}} = B\frac{w_t - w_s}{(t - s)^{1/2}} \,. \qquad (6.10)$$

From the properties of the Wiener process it follows that

$$E\left[B\frac{w_\tau - w_t}{(\tau - t)^{1/2}}\right] = E\left[B\frac{w_t - w_s}{(t - s)^{1/2}}\right] = 0$$

$$E\left[B\frac{w_\tau - w_t}{(\tau - t)^{1/2}}\right]^2 = E\left[B\frac{w_t - w_s}{(t - s)^{1/2}}\right]^2 = B^2$$

$$E\left[B\frac{w_\tau - w_t}{(\tau - t)^{1/2}} \times B\frac{w_t - w_s}{(t - s)^{1/2}}\right] = 0 \,.$$

Since the terms on the left-hand side of equation (6.10) are independent, we have a Gaussian distribution with zero mean and covariance B^2. As we consider X_m a set of given random-process realizations for given Θ^i, where $i = 1, 2, \ldots, m$, which satisfy equation (6.9), then each trajectory from X_m must possess these properties also. This means that for the GI trajectory of each ith-patient, where $i = 1, \ldots, m$, the term

$$z_{ij}(\lambda) = \frac{(\ln x_j^i)/x_{j-1}^i + \lambda(t_j^i - t_{j-1}^i)}{(t_j^i - t_{j-1}^i)^{1/2}} \,,$$

where $x_j^i = x_{t_j}^i$, $j = 1, \ldots, n_i$, has a Gaussian distribution $N(0, B^2)$ and z_{ij} and z_{ik} are independent for $j \neq k$. The independence of these terms, calculated using different trajectories, is obvious. These properties are the basis for the application of the maximum-likelihood method for estimating parameters λ and B.

So, we have a sample of Σn_i, where $i = 1, \ldots, m$, of independent values given by

$$Z = (z_{11}, z_{12}, \ldots, z_{1n_1}, z_{21}, z_{22}, \ldots, z_{2n_2}, \ldots, z_{m1}, z_{m2}, \ldots, z_{mn_m}) \,.$$

The distribution density of Z has the form

$$p(z, \lambda, B) = \frac{1}{(2\pi)^{1/2}B} \exp\left[-\frac{z^2(\lambda)}{2B^2}\right] .$$

This result allows us to use the maximum-likelihood principle for parameter estimation. As long as the values included in Z are independent, then for conditional joint density $p(Z|\alpha)$ we have the expression

$$p(Z|\lambda, B) = \prod_{i=1}^{m} \prod_{j=1}^{n_i} p(z_{ij}(\lambda), \lambda, B) .$$

Then

$$l(\lambda, B) = \ln p(Z|\lambda, B)$$
$$= -\frac{1}{2}\left[\sum_{i,j} \ln(2\pi B^2) + \sum_{i,j} \frac{z_{ij}^2(\lambda)}{B^2}\right] .$$

The search for this function's maximum with respect to λ and B^2 is equivalent to the following problem:

$$\min_{\lambda, B^2}\{L(\lambda, B^2) = -2l(\lambda, B^2) - \sum_{i,j} \ln 2\pi\} .$$

The estimates of the maximum likelihood λ_m and B_m^2 are defined from the condition

$$L(\lambda, B^2) = N \ln B^2 + \frac{1}{B^2} \sum_{i=1}^{m} \sum_{j=1}^{n_i} \frac{[(\ln x_j^i)/x_{j-1}^i + \lambda(t_j^i - t_{j-1}^i)]^2}{t_j^i - t_{j-1}^i}$$
$$\Rightarrow \min_{\lambda, B^2} ,$$

where $N = \Sigma n_i$, $i = 1, \ldots, m$: that is, from the equations

$$\frac{\partial}{\partial \lambda} L(\lambda, B^2) = 0$$
$$\frac{\partial}{\partial B^2} L(\lambda, B^2) = 0 .$$

It is simple to show that this problem has a unique solution that can be written in the following form:

$$\lambda_m = \frac{1}{T_m} \sum_{i,j} \ln\left(\frac{x_{j-1}^i}{x_j^i}\right)$$

$$B_m^2 = \frac{1}{N} \sum_{i,j} \left[\frac{(\ln x_{j-1}^i)/x_j^i}{(t_j^i - t_{j-1}^i)^{1/2}} - \lambda_m \left(t_j^i - t_{j-1}^i\right)^{1/2}\right]^2 , \qquad (6.11)$$

where

$$T_m = \sum_{i,j}(t^i_j - t^i_{j-1}) = \sum_{i=1}^{m}(t^i_{n_i} - t^i_0)$$

is the total observation time.

Now we need to use the following theorem so first we show that it is valid.

Theorem 6.2

The random values

$$\frac{NB^2_m}{B^2} \quad \text{and} \quad [(N-1)/NT_m]^{1/2}\frac{(\lambda_m - \lambda)}{B_m}$$

have a χ^2 and Student S_{N-1} distribution respectively, where λ and B are unknown true values.

Proof

Let us introduce one index into equation (6.11) and denote

$$y_k = \frac{(\ln x^i_{j-1})/x^i_j}{(\Delta t_k)^{1/2}} - \lambda(\Delta t_k)^{1/2}$$

$$\Delta t_k = t^i_j - t^i_{j-1}$$

$$i = 1,\ldots,m, \quad j = 1,\ldots,n_i$$

$$k = \sum_{l=0}^{i-1} n_l + j, \quad n_0 = 0 \ .$$

The random terms y_k are independent and Gaussian, with zero mean and covariance B^2. Keeping this in mind, the equations in (6.11) are represented in the following form:

$$(\lambda_m - \lambda)(T_m)^{1/2} = \frac{1}{(T_m)^{1/2}}\sum_{i=1}^{m}(\Delta t_i)^{1/2}y_i$$

$$NB^2_m = \sum_{i=1}^{N} y^2_i - (\lambda_m - \lambda)^2 T_m \ .$$

Using the formulae for λ, we obtain

$$(\lambda_m - \lambda)(T_m)^{1/2} = \sum_{i=1}^{N}\alpha_i y_i, \quad \alpha_i = (\Delta t_i/T_m)^{1/2}$$

$$NB^2_m = \sum_{i=1}^{N} y^2_i - \left(\sum_{i=1}^{N}\alpha_i y_i\right)^2 \ . \tag{6.12}$$

Let A be an orthogonal matrix, that is, $AA^T = I$. Introduce an N-dimensional, random term $u = Ay$. Then

$$Euu^T = EAyy^T A^T = A\left(Eyy^T\right)A^T$$

and therefore the terms $u^1, u^2, \ldots, u^N$ consisting of a vector column are independent, and the covariance of each of them is equal to B^2. Let the elements of the last row a_N of matrix A be determined by the expression

$$a_N^i = (\Delta t_i/T_m)^{1/2} \ ,$$

where $i = 1, \ldots, N$. Note that $a_N a_N^T = 1$. Then equation (6.12) can be written using $u^1, u^2, \ldots, u^N$:

$$(\lambda_m - \lambda)(T_m)^{1/2} = u_N$$

$$NB_m^2 = \sum_{i=1}^{N}(u^i)^2 - (u^N)^2 = \sum_{i=1}^{N-1}(u^i)^2 \ . \qquad (6.13)$$

From (6.13) it follows that the terms $(\lambda_m - \lambda)(T_m)^{1/2}$ and NB_m^2 are independent and

$$N\left(\frac{B_m^2}{B^2}\right) = \sum_{i=1}^{N-1}\frac{(u^i)^2}{B^2} \qquad (6.14)$$

represents the sum of normally distributed random variables squared. So, NB_m^2/B^2 has a χ_{N-1}^2 distribution with $N-1$ degrees of freedom. It is known (Cramer, 1946) that the ratio $\zeta(\xi/n)^{-1/2}$, where ζ and ξ are independent, $\zeta \sim N(0,1)$ and $\xi \sim \chi_n^2$, has a Student distribution with n degrees of freedom. Therefore, the random value

$$\left[(\lambda_m - \lambda)(T_m)^{1/2}/B\right]\left\{NB_m^2/[B^2(N-1)]\right\}^{-1/2}$$

$$= \frac{\lambda_m - \lambda}{B_m}\left[T_m\,\frac{(N-1)}{N}\right]^{1/2}$$

has a Student distribution S_{N-1}.

Note that in expression (6.11) for B_m^2, division by N must be replaced with division by $N-1$, since from equation (6.14) we obtain

$$EB_m^2 = \frac{1}{N}\sum_{i=1}^{N-1}E(u^i)^2 = \frac{N-1}{N}B^2 \ ,$$

that is, the estimate B_m^2 is asymptotically unbiased. If we make the noted replacement, then $E B_m^2 = B^2$. For a derivation of confidence boundaries let us assume that the distributions for the following two terms

$$\beta = (\lambda_m - \lambda)(T_m)^{1/2}/B_m$$
$$\eta = (N-1)B_m^2/B^2$$

are known. Let p be close to probability 1: for example, $p = 0.90 / 0.95$. Define a $p100\%$ confidence interval for λ, that is, λ_1, λ_2 such that

$$P(\lambda_1 \leq \lambda \leq \lambda_2) = p .$$

According to Student distribution tables for the number of degrees of freedom, $N-1$, one can find the value β_p at which

$$P(|\beta| \leq \beta_p) = p ,$$

that is, with probability p

$$|\lambda_m - \lambda|(T_m)^{1/2}/B_m \leq \beta_p ,$$

from which

$$\lambda_{1,2} = \lambda_m \pm \beta_p \left[\frac{B_m}{(T_m)^{1/2}} \right] .$$

The accuracy of the parameter estimate is higher when the variance of the experimental data is smaller and the total observation time is larger. The χ^2 distribution is not symmetrical, unlike the Student one. For β_1 and β_2 such that $P(\beta_1^2 \leq \beta^2 \leq \beta_2^2) = p$, we find at first two values, χ_1 and χ_2, from the conditions

$$P(\eta < \chi_1) = (1-p)/2$$
$$P(\eta > \chi_2) = (1-p)/2, \ \chi_1 < \chi_2 .$$

This means that

$$P(\chi_1 \leq \eta \leq \chi_2) = 1 - P(\eta < \chi_1) - P(\eta > \chi_2) = p .$$

Then

$$P(\chi_1 \leq (N-1) B_m^2/B^2 \leq \chi_2) = p ,$$

or

$$P((N-1)B_m^2/\chi_2 \leq B^2 \leq (N-1)B_m^2/\chi_1) = p .$$

Table 6.1. Results of calculations based on treatment with drug A or drug B.

	$P(\lambda_1 < \lambda < \lambda_2)=0.95$			$P(B_1^2 < B^2 < B_2^2)=0.95$		
	λ_1	λ_m	λ_2	B_1^2	B_m^2	B_2^2
Drug A	0.168	0.174	0.179	0.031	0.045	0.077
Drug B	0.126	0.130	0.135	0.035	0.046	0.069
No drug	0.120	0.123	0.128	0.028	0.037	0.054

Therefore,

$$B_1^2 = (N-1)B_m^2/\chi_2$$
$$B_2^2 = (N-1)B_m^2/\chi_1 \; ,$$

where $P(\eta > \chi_1) = (1+p)/2, P(\eta > \chi_2) = (1-p)/2, \eta \sim \chi_{N-1}^2$.

We should now focus attention on the following important conclusion. The description of the random character of real data by means of the stochastic model allows us to construct confidence intervals for parameters needed to solve the practical problems. Let us illustrate one case next.

6.5 Effectiveness of Alternative Therapies

In *Figures 6.2* and *6.3* the trajectories of the GI in two groups of patients with viral hepatitis treated with different drugs A and B are presented. The results of calculations based on the two given groups are presented in *Table 6.1*. From the table, we can see a significant improvement in recovery rate for the control group corresponding to drug A, which is more favorable than drug B. It is interesting that, within the given limits of accuracy, a difference in the values of the B parameters corresponding to different groups is not observed. This may be explained by the action of factors that are not included in the model, but which are random phenomena for the considered groups. As far as these group factors are concerned, we cannot observe any significant difference in the intensity parameter B^2.

6.6 Inspection of a Disease Course

Following the methods proposed by Pogozhev (1988), let us use the set, X_m, of GI trajectories corresponding to the group of patients with favorable disease dynamics; from this we can construct confidence intervals

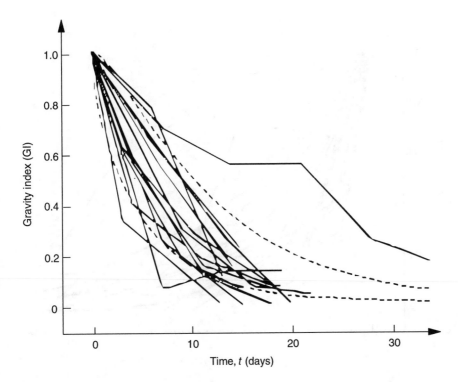

Figure 6.2. Trajectories of GI changes for patients treated with drug A.

for the normal course of a disease. To this end, we need to construct a representative sample, X_m, of trajectories to find boundaries X_t^+ and X_t^- such that

$$P\{X_t^- \le X_t \le X_t^+\} = p \ ,$$

where $p = 0.9$ or 0.95.

Let λ_m and B_m be calculated by the set X_m, and let m be sufficiently large to consider $\lambda_m = \lambda^*$ and $B_m = B^*$. From the model we have

$$\zeta = \frac{\ln x_t + \lambda^* t}{B^*(t)^{1/2}} = \frac{w_t}{(t)^{1/2}} \ ,$$

where $E\zeta = 0, E\zeta^2 = 1$, and ζ has a Gaussian distribution. Let us define q_p such that

$$P\{|\zeta| \le q_p\} = p \ .$$

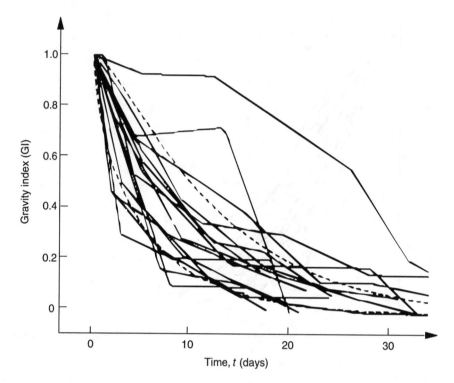

Figure 6.3. Trajectories of GI changes for patients treated with drug B.

Then,

$$P\left\{\left|\frac{\ln x_t + \lambda^* t}{B^*(t)^{1/2}}\right| \le q_p\right\} = p \ .$$

Thus,

$$x_t^+ = \exp\{-[\lambda^* t - q_p B^*(t)^{1/2}]\}$$
$$x_t^- = \exp\{-[\lambda^* t + q_p B^*(t)^{1/2}]\} \ .$$

This confidence region is shown in *Figure 6.4*. The construction of a confidence region is very important for the analysis of the dynamics of a disease because this allows one to control the state of an individual patient and to make appropriate, timely decisions concerning further strategies for treating the patient.

Let the individual value of the GI x_t at $t > 0$ lie outside the confidence region. The probability of this happening is $1 - p = 0.05$. This

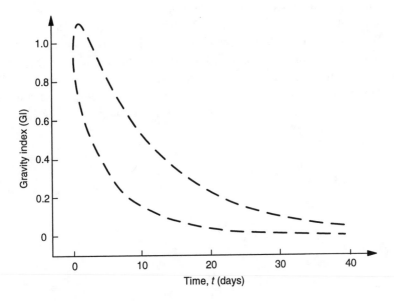

Figure 6.4. A confidence interval for a normal process of rehabilitation of an afflicted organism.

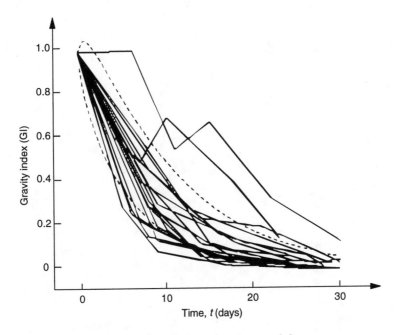

Figure 6.5. Simulation of real data by the model.

deviation cannot be explained by random factors. It is necessary to find the cause of this deviation and to eliminate it by treatment.

To conclude this chapter, we illustrate the correspondence between the stochastic model of equation (6.8) and clinical data. *Figure 6.5* presents results which imitate the actual trajectories (*Figure 6.1*) by those given by the model in equation (6.8). The correspondence of this model to real data, using statistical criteria, is given by Zuev (1988).

Chapter 7

Statistical Parameter Estimation

In this chapter the parameter estimation method, which was studied in Chapter 6, is generalized for a more complex model. We consider the case when the model is represented by the following system of differential equations:

$$\frac{d}{dt}x_t = f(x_t, \alpha), \ \ t \in [0, T], \ \ x_0 = c \ . \tag{7.1}$$

Here, $x_t \in \mathbb{R}^n$ is a vector of state variables and $\alpha \in \mathbb{R}^L$ is a vector of parameters. The problem under consideration is how to estimate α by using sample X_m values (see Chapter 6). First of all, we describe the random character of the real trajectories in the framework of the model given by equation (7.1).

7.1 Stochastic Model for Data Description

As a start, let us suppose that random error is absent and that the stochastic character of the trajectories is completely explained by those factors which are not incorporated in the model. By changing the coefficients, it is possible to reproduce the observable dynamics of state variables. Assume that for each trajectory $x_t^i \in X_m$ $(i = 1, 2, \ldots, m)$ there is a corresponding piecewise continuous function $\{\alpha_t^i, t \in [0, T]\}$ such that

$$x_t(\alpha_t^i) = x_t^i, \ \ t \in \Theta, \ \ x_0(\alpha^i) = c \ ,$$

119

where $x_t(\alpha_t^i)$ is the solution of equation (7.1) when we replace α by α_t^i. We can consider the set of functions $\{\alpha_t^i, t \in [0,T], i = 1,2,\ldots,m\}$ as the set of realizations of some stochastic process given by $\alpha_t = \{\alpha_t(\omega), \omega \in \Omega, t \in [0,T]\}$ defined in $\mathbb{R}^l$ such that

$$E\alpha_t = \int_\Omega \alpha_t(\omega)dP(\omega) = \alpha^*, \ \forall t \in [0,T] \ , \tag{7.2}$$

since the regular differences between trajectories are absent and the coefficients of the model in (7.1) are constants.

So, the set X_m represents the realizations of the stochastic process $\{\tilde{x}_t(\omega), \omega \in \Omega, t \in [0,T]\}$ on Θ. This process is described by the following equation:

$$\begin{aligned} \frac{d}{dt}x_t &= f(x_t, \alpha_t) \\ x_0 &= c, \ t \in [0,T] \ . \end{aligned} \tag{7.3}$$

As far as X_m is concerned, we assume that all trajectories fit the general function $x_t(\alpha^*)$. However, the validity of condition (7.2) is not sufficient for solutions of the model in equation (7.3) to have the same property. Let the process α_t be such that condition (7.2) is fulfilled and for each $\omega \in \Omega$ and any $t \in [0,T]$ we have $\alpha_t(\omega) = \alpha(\omega)$. In this case, the set of solutions given by equation (7.3), which corresponds to different $\alpha(\omega)$, can even consist of qualitatively different solutions. Such a situation can be excluded if we assume that, except for condition (7.2), the following approximate equality takes place:

$$\frac{1}{T}\int_0^T \alpha_t dt \approx \alpha^* \ . \tag{7.4}$$

The exact equality (in the sense of probability convergence) takes place when $T \to \infty$ if the dependence between α_t and $\alpha_{t+\tau}$ is weakened with increasing τ.

For a formal statement of this fact we represent the process α_t in the following way: $\alpha_t = \alpha^* + \xi_{t/\varepsilon}, \xi_t \in \mathbb{R}^L$ where ε is a parameter which has a small value ($\varepsilon > 0$), $E\xi_t = 0$, $E|\xi_t|^2 < \infty$, and $\forall t \in [0,T]$. Then, according to condition (7.4), we assume that for any $\delta > 0$ and $T > 0$

$$\lim_{\varepsilon \to 0} P\left\{\left|\frac{1}{T}\int_t^{t+T} \xi_{s/\varepsilon}ds\right| > \delta\right\} = 0 \tag{7.5}$$

uniformly on t. So the model in equation (7.3) acquires the form

$$\frac{d}{dt}x_t = f(x_t, \alpha^* + \xi_{t/\varepsilon})$$
$$x_0 = c, \quad t \in [0, T] . \tag{7.6}$$

Denote x_t^ε as a solution of this equation, which we consider as the model of the real process. The small parameter on the right-hand side of equation (7.6) means that the random variable is fast. In order to clarify this fact we change the time-scale, i.e., make the replacement $z_{t/\varepsilon} = x_t^\varepsilon$. From equation (7.6) we have

$$\frac{d}{dt}z_{t/\varepsilon}\frac{1}{\varepsilon} = \frac{d}{dt}x_t^\varepsilon = f(x_t^\varepsilon, \alpha^* + \xi_{t/\varepsilon}) = f(z_{t/\varepsilon}, \alpha^* + \xi_{t/\varepsilon}).$$

For the new variable we have the equation

$$\frac{d}{dt}z_t = \varepsilon f(z_t, \alpha^* + \xi_t)$$

$$t \in [0, T/\varepsilon] .$$

Thus z_t is a slow variable with respect to ξ_t.

The stochastic system of equation (7.6) and the limit relation in equation (7.5) represent the formal description of the correspondence between the deterministic model in equation (7.1) and real trajectories of state variables. The solution disturbed by random processes given by the model in equation (7.6) is a fast, random fluctuation along the solution $x_t(\alpha^*)$. This is a result of the following theorem.

Theorem 7.1
Let the random process ξ_t in the system in equation (7.6) be satisfied by condition (7.5) and

$$\sup_{0 \le t \le T} E|f(x, \xi_t)|^2 < \infty .$$

Then, for all $T > 0$, $q > 0$

$$\lim_{\varepsilon \to 0} P\left\{ \sup_{0 \le t \le T} |x_t^\varepsilon - x_t(\alpha^*)| > q \right\} = 0 .$$

Here x_t^ε means the solution of equation (7.6) and $x_t(\alpha^*)$ is a solution of equation (7.1). This means that at small values of ε with probability close to 1 the trajectories of the disturbed system are situated within a small range of the solution $x_t(\alpha^*)$. Their random deviations are near this solution.

So, when we consider trajectories of the real process as solutions of the model in equation (7.6) we avoid a discrepancy between modeled and observed data. However, this is not sufficient for the solution of the estimation problem because we do not know anything about the distribution of the deviations $x_t^\varepsilon - x_t(\alpha)$, where $t \in \Theta$. In order to be able to solve the problem, we take into consideration the fact that the term ε has a characteristic time of random oscillations of the order of $x_t(\alpha^*)$ (hours) compared with the duration of the process in the system (days). So we can consider ε as having a sufficiently small value for us to be able to use the properties of limited distribution ($\varepsilon \to 0$) of the difference $x_t^\varepsilon - x_t(\alpha^*)$ for the solution of the estimation problem. Keeping in mind the assumptions made in Section 7.1 about the right-hand sides of equation (7.6) we can now formulate the simplified variant of the theorem which was proved by Ventcel and Freidlin (1975).

Theorem 7.2

Let the components of the vector function $f(x, y)$ have, in all space, the first and second partial derivatives. The random process with values in $\mathbb{R}^L$ is assumed to have piecewise continuous (with probability 1) trajectories, and to satisfy the condition of strong mixing with coefficient $\beta(\tau)$ such that:

$$\int_0^\infty \tau(\beta(\tau))^{1/5}d\tau < \infty$$

$$\sup_{x,t} E|f(x, \xi_t)|^3 < M < \infty \ .$$

Then, as $\varepsilon \to 0$, the process

$$v_t^\varepsilon = (x_t^\varepsilon - x_t(\alpha^*))/(\varepsilon)^{1/2}$$

converges weakly on the interval [0,T] to a Markovian process that satisfies the system of linear differential equations

$$\frac{d}{dt}v_t^0 = f_x(x_t(\alpha^*), \alpha^*)v_t^0 + \frac{d}{dt}w_t, \quad v_0^0 = 0 \ .$$

Here, w_t is a Gaussian process with independent increments, zero mathematical expectation, and a covariance matrix $R_t \doteq (R_t^{ij})$ given by

$$R_t^{ij} \quad = \quad Ew_t^i w_t^j = \int_0^t Q^{ij}(x_s(\alpha^*))ds$$

$$Qij(x) = \lim_{T\to\infty} \frac{1}{T} \int_0^T \int_0^T Q_1^{ij}(x,s,t)dsdt$$

$$Q_1^{ij}(x,s,t) = Ef^i(x,\xi_t)f^j(x,\xi_s) \tag{7.7}$$

where i and $j = 1,2,\ldots,n$, $f^i(x,y)$ are the components of the vector function $f(x,y)$, and $f_x(x,y)$ is a matrix of derivatives with respect to x.

So we obtain the result that at certain specific times in the lifetimes of the variables considered the deviation process given by the equation $u_t = x_t^\varepsilon - x_t(\alpha^*)$ can be approximately considered as a Markovian random process which is satisfied by the following equation:

$$\frac{d}{dt}u_t = f_x(x_t(\alpha^*),\alpha^*)u_t + (\varepsilon)^{1/2}\frac{d}{dt}w_t, \quad u_0 = 0 \ . \tag{7.8}$$

This means that for $t_0 < t_1 < \cdots < t_N$, the joint density distribution

$$p(u_{t_i}, i = 1,2,\ldots,N) = p(u_{t_0},u_{t_1},\ldots,u_{t_N})$$

is represented in the form of multiplications of conditional densities given by

$$p(u_{t_i}, i = 1,2,\ldots,N) = \prod_{i=1}^{N} p(u_{t_i}|u_{t_{i-1}}) \ ,$$

where $p(x|y)$ is a conditional Gaussian density function at fixed y. Since the Gaussian random variable is completely defined by the mathematical expectation and the covariance matrix, we need only write the equation for their definition. For mathematical expectation

$$\mu_t(\alpha) = E_\alpha u_t = \int_{-\infty}^{\infty} (x_t - x_t(\alpha))p(x_t|\alpha)dx_t, \quad \mu_t(\alpha) \in \mathbb{R}^n \ .$$

From equation (7.8) one can find

$$\frac{d}{dt}\mu_t(\alpha) = f_x(x_t(\alpha),\alpha)\mu_t(\alpha)$$

where $\mu_0(\alpha) = 0$ and $t \in [0,T]$. It follows from this that $\mu_t(\alpha) = E_\alpha u_t = 0, \forall t \in [0,T]$, $\alpha \in D$, and then $E_\alpha x_t^\varepsilon = x(t,\alpha)$: that is, on average, our model describes the real process. For the conditional mathematical expectation $\mu_{ts}(\alpha) = E_\alpha(u_t|u_s)$, we obtain

$$\frac{d}{dt}\mu_{\tau s}(\alpha) = f_x(x_\tau(\alpha),\alpha)\mu_{\tau s}(\alpha) \ ,$$

where $\mu_{ss} = u_s$ and $\tau \in [s, t]$. It should be noted that, according to equation (7.10), for small values of ε we can (with probability close to 1) consider that

$$|u_s| = |x_s^\varepsilon - x_s(\alpha^*)| = |\delta_s| < \sigma \ ,$$

where σ has a small, positive value. Then, from the proposition about the stability of the solution $x_t(\alpha)$, according to a linear approximation it follows that

$$\mu_{ts}(\alpha) = \delta x_t = x_t(\alpha, s, x_s^\varepsilon) - x_t(\alpha, s, x_s(\alpha)) \to 0$$

when $t \to \infty$, $\delta x_s = x_s^\varepsilon - x_s(\alpha)$.

Therefore, when $t \gg s$, in practice we have $\mu_{ts}(\alpha) = \mu_t(\alpha) = 0$, and for mathematical expectation we have

$$E_\alpha(x_t^\varepsilon | x_s^\varepsilon) = E_\alpha(x_t^\varepsilon) = x_t(\alpha) \ .$$

This is an illustration of the fact that the process x_t^ε eventually "forgets" its own previous states. In order to obtain the equations which describe the changes in covariance, we consider the variable $z_t = u_t^i u_t^j$, where u_t^i and u_t^j are the components of vector u_t. Expanding z_t in a Taylor series for the point $(t + \Delta t)$, and omitting terms of order greater than two, we obtain

$$
\begin{aligned}
z_{t+\Delta t} &= z_t + \frac{\partial z_t}{\partial u^i} \Delta u_t^i + \frac{\partial z_t}{\partial u^j} \Delta u_t^j + \frac{\partial^2 z_t}{\partial u^i \partial u^j} \Delta u^i \Delta u^j \\
&= z_t + u_t^j \Delta u_t^i + u_t^i \Delta u_t^j + \Delta u_t^i \Delta u_t^j \ .
\end{aligned}
$$

Taking into account equation (7.8), we have

$$
\begin{aligned}
\Delta z_t &= u_t^j \sum_k f^{ik} u_t^k \Delta t + u_t^i \sum_k f^{jk} u_t^k \Delta t + u_t^i \Delta w_t^j \sqrt{\varepsilon} \\
&\quad + \Delta w_t^i \Delta w_t^j \varepsilon + (\Delta t)^2 \sum_k f^{ik} u_t^k \sum_k f^{jk} u_t^k \\
&\quad + \sqrt{\varepsilon} \Delta t \Delta w_t^i \sum_k f^{jk} u_t^k + \sqrt{\varepsilon} \Delta t \Delta w_t^j \sum_k f^{ik} u_t^k \ ,
\end{aligned}
$$

where f^{ij} is an element of matrix $f_x(x_t(\alpha), \alpha)$. Then

$$
\begin{aligned}
E \Delta z_t &= \left(\sum_k f^{ik} K_t^{jk} + \sum_k f^{jk} K_t^{ik} \right) \Delta t + \varepsilon E \Delta w_t^i \Delta w_t^j \\
&\quad + (\Delta t)^2 E \sum_k f^{jk} u_t^k \sum_k f^{ik} u_t^k \ ,
\end{aligned}
\tag{7.9}
$$

where $K_t^{ij} = Eu_t^i u_t^j$. Since

$$
\begin{aligned}
\Delta w_t^i \Delta w_t^j &= (w_{t+\Delta t}^i - w_{t_i}^i)(w_{t+\Delta t}^j - w_t^j) \\
&= w_{t+\Delta t}^i w_{t+\Delta t}^j - w_t^i w_t^j - (w_{t+\Delta t}^i - w_t^i)w_t^j \\
&\quad -(w_{t+\Delta t}^j - w_t^j)w_t^i \;,
\end{aligned}
$$

and taking into account that w_t is an independent-increment process, we obtain

$$
E\Delta w_t^i \Delta w_t^j = Ew_{t+\Delta t}^i w_{t+\Delta t}^j - Ew_t^i w_t^j \;.
$$

From this and equation (7.7),

$$
E\Delta w_t^i \Delta w_t^j = \int_t^{t+\Delta t} Q^{ij}(x_s(\alpha))ds \;.
$$

Then, dividing both sides of equation (7.9) by Δt, and in the limit taking Δt to zero, we obtain the equation for K^{ij}:

$$
\frac{d}{dt}K_t^{ij} = \frac{d}{dt}Ez_t = \sum_r f_x^{ir} K_t^{jr} + \sum_r f_x^{jr} K_t^{ir} + \varepsilon Q^{ij}(x_t(\alpha)) \;,
$$

or in matrix form

$$
\frac{d}{dt}K_t = f_x(x_t(\alpha), \alpha)K_t + K_t f_x^T(x_t(\alpha), \alpha) + \varepsilon Q(x_t(\alpha)) \;. \tag{7.10}
$$

For the determination of $Q(x_t(\alpha))$ on the right-hand side of equation (7.10) we use the representation (see section 7.1)

$$
f(x, \alpha) = F(x)\alpha
$$

and, according to equation (7.7),

$$
\begin{aligned}
Q_1^{ij}(x, s, t) &= Ef^i(x, \xi_t)f^j(x, \xi_s) = E\sum_k F^{ik}(x)\xi_t^k \sum_k F^{jk}(x)\xi_s^k \\
&= \sum_{r,k} F^{ik}(x)F^{jr}(x)E\xi_t^r \xi_s^k \\
Q^{ij}(x) &= \lim_{T\to\infty} \frac{1}{T}\int_0^T \int_0^T \sum_{r,k} F^{ir}(x)F^{jk}(x)E\,\xi_t^r \xi_s^k \, dt\, ds \\
&= \sum_{r,k} F^{ir}(x)F^{jk}(x)\lim_{T\to\infty} \frac{1}{T}\int_0^T \int_0^T E\,\xi_t^k \xi_s^r \, ds\, dt \\
&= \sum_{r,k} F^{ir}(x)F^{jk}(x)g^{kr} \;.
\end{aligned}
$$

In the last expression

$$g^{kr} = \lim_{T \to \infty} \int_0^T \int_0^T E\, \xi_t^k \xi_s^r \, ds\, dt \ . \tag{7.11}$$

The elements of matrix G, which is given in equation (7.11), are intensities of random perturbation. Since in equation (7.10) the matrix $Q(x)$ is multiplied by ε, we denote $\Gamma = \varepsilon G$. The matrix Γ is called the intensity matrix. Further, we let $\varepsilon Q(x_t) = Q(x_t, \Gamma)$. For the conditional covariance matrix $K_{ts} = \text{cov}(u_t, u_t/u_s)$, we have the equation

$$\frac{d}{dt} K_{\tau s} = f_x(x_\tau(\alpha), \alpha) K_{\tau s} + K_{\tau s} f_x^T(x_\tau(\alpha), \alpha) + Q(x_\tau(\alpha), \Gamma) \ ,$$

where $K_{ss} = 0$ and $\tau \in [s, t]$. The elements of matrix Γ are unknown and need to be estimated by using sample trajectories and the vector α.

7.2   The Maximum-likelihood Solution

There is a sample of independent realizations

$$X_m = \{X^i, i = 1, 2, \ldots, m\}, \quad X^i = \{x_t^i, t \in \Theta\}$$

of the random process

$$\tilde{X}_t = \{\tilde{x}_t(\omega), \quad t \in [0, T], \quad \omega \in \Omega\}$$

with known distribution [since it is described by the model in equation (7.6)]. According to the maximum-likelihood principle, we can find the estimate α_m of unknown vector α^* from the condition

$$\max_{\substack{\alpha \in D \\ \Gamma \in R_+^{l \times l}}} \ln p(X_m | \alpha, \Gamma) = \ln p(X_m | \alpha_m, \Gamma_m) \ .$$

From the independence of trajectories it follows that

$$p(X_m | \alpha, \Gamma) = \prod_{i=1}^m p(X^i | \alpha, \Gamma) \ ,$$

where

$$p(X | \alpha, \Gamma) = p(x_{t_0}, x_{t_1}, \ldots, x_{t_N}; t_0, t_1, \ldots, t_N) \tag{7.12}$$

is an N-dimensional probability density function of the random process x_t^ε. This function gives the joint distribution of random-process values at the moments $t_0, t_1, \ldots, t_N$. The likelihood function is given by the following expression:

$$\Phi_m(\alpha, \Gamma) = \frac{1}{m} \ln p(X_m|\alpha, \Gamma) = \frac{1}{m} \sum_{i=1}^{m} \ln p(X^i|\alpha, \Gamma) \ .$$

Then

$$\max_{\alpha, \Gamma} \Phi_m(\alpha, \Gamma) = \Phi_m(\alpha_m, \Gamma_m) \ .$$

According to the results of the above section, the process of random deviations is Markovian. Consequently,

$$p(X|\alpha, \Gamma) = \prod_{i=1}^{N} p(u_{t_i}|u_{t_{i-1}}) \ .$$

Then expression (7.12) takes the form

$$\Phi_m(\alpha, \Gamma) = \frac{1}{m} \sum_{j=1}^{m} \sum_{i=1}^{N} \ln p(x_i^j - x_i(\alpha)|x_{i-1}^j - x_{i-1}(\alpha)) \ ,$$

where $x_i^j = x_{t_i}^j \in X^j$ and $x_i(\alpha) = x_{t_i}(\alpha)$.

Since $u_t = \tilde{x}_t - x_t(\alpha)$ is a Gaussian process, we have

$$
\begin{aligned}
p(x_i - x_i(\alpha)|x_{i-1} - x_{i-1}(\alpha)) \ = \ & ((2\pi)^n \det(K_{i,i-1}))^{-1/2} \\
& \times \exp\Big[-\frac{1}{2}(x_i - x_i(\alpha) - \mu_{i,i-1})^T \\
& \times K_{i,i-1}^{-1}(x_i - x_i(\alpha) - \mu_{i,i-1})\Big] \ ,
\end{aligned}
$$

where $K_{i,i-1} = K_{t_i t_{i-1}}(\alpha), \mu_{i,i-1} = \mu_{t_i t_{i-1}}(\alpha)$, and $\mu_{i,i-1} \in \mathbf{R}^n$.

The distribution moments are given by the following equations:

$$
\begin{aligned}
\frac{d}{dt}\mu_{tt_{i-1}}(\alpha) \ = \ & f_x(x_t(\alpha), \alpha)\mu_{tt_{i-1}} \\
\frac{d}{dt}K_{tt_{i-1}}(\alpha) \ = \ & f_x(x_t(\alpha), \alpha)K_{tt_{i-1}}(\alpha) \\
& + K_{tt_{i-1}}(\alpha)f_x^T(x_t(\alpha), \alpha) + Q(x_t(\alpha), \Gamma) \ , \quad (7.13)
\end{aligned}
$$

where $t \in [t_{i-1}, t_i], \mu_{t_{i-1}t_{i-1}}(\alpha) = x_{i-1} - x_{i-1}(\alpha)$, and $K_{t_{i-1}t_{i-1}}(\alpha) = 0$. Note that each trajectory X^j from X_m can be given on its own set Θ^j. If x_t^i and x_t^j at $t \neq s$, where $i, j = 1, 2, \ldots, m$, are independent according to the conditions of an experiment, then for every trajectory

$$p(x_k^i - x_k^i(\alpha)|x_{k-1}^j - x_{k-1}^j(\alpha)) = p(x_k^i - x_k^i(\alpha)) \ ,$$

where $i, j = 1, 2, \ldots, n$ and $k = 1, 2, \ldots, N$, and

$$
\begin{aligned}
p(x_i - x_i(\alpha)) \;=\; & ((2\pi)^n \det(K_i))^{-1/2} \\
& \times \exp\left[-\frac{1}{2}(x_i - x_i(\alpha))^T K_i^{-1}(x_i - x_i(\alpha))\right] ,
\end{aligned}
$$

where $K_i = K_{t_i}(\alpha)$.

Thus, for any $t \in [0, T]$,

$$
\frac{d}{dt} K_t(\alpha) = f_x(x_t(\alpha), \alpha) K_t(\alpha) + K_t(\alpha) f_x^T(x_t(\alpha), \alpha) + Q(x_t(\alpha), \Gamma)
$$

where $t \in [0, T]$ and $K_0 = 0$. Then the likelihood function has the following expression:

$$
\begin{aligned}
\Phi_m(\alpha, \Gamma) \;=\; & \sum_{t \in \Theta} \frac{1}{m} \sum_{i=1}^{m} \left[-\frac{1}{2} \ln((2\pi)^n \det(K_t)) \right. \\
& \left. -\frac{1}{2}(x_t^i - x_t(\alpha))^T K_t^{-1}(x_t^i - x_t(\alpha)) \right] \\
\;=\; & -\frac{1}{2} \sum_{t \in \Theta} \frac{1}{m} \sum_{i=1}^{m} \ln((2\pi)^n \det(K_t)) \\
& -\frac{1}{2} \sum_{t \in \Theta} \frac{1}{m} \sum_{i=1}^{m} (x_t^i - x_t(\alpha))^T K_t^{-1}(x_t^i - x_t(\alpha)) . \quad (7.14)
\end{aligned}
$$

If the matrix K_t is known for all $t \in \Theta$, then from equation (7.14) the criterion of the least-squares method is obtained as

$$
\sum_{t \in \Theta} \frac{1}{m} \sum_{i=1}^{m} (x_t^i - x_t(\alpha))^T K_t^{-1}(x_t^i - x_t(\alpha)) \to \min_{\alpha \in D \subset R^L} .
$$

The effective estimate of a vector α is given by the least-squares method only in the case when the covariance matrix is known for all $t \in \Theta$. So it should be noted that a criterion which is often used in practice,

$$
\sum_{t \in \Theta} \frac{1}{m} \sum_{i=1}^{m} (x_t^i - x_t(\alpha))^T (x_t^i - x_t(\alpha)) \to \min_{\alpha \in D \subset R^L}
$$

can provide an effective estimate if, and only if, $K_t = I \; \forall t \in \Theta$, and values of the real process at time t are independent.

7.3 Convergence of Estimates

The asymptotic properties of the maximum-likelihood estimates are well known (Cramer, 1946; Mudrov and Kushko, 1983). Here, we deal with a sample of realizations of the random process. First of all, we note that the likelihood function $\Phi_m(\alpha)$, given by

$$\Phi_m(\alpha) = \frac{1}{m} \sum_{i=1}^{m} \ln p(X^i|\alpha)$$

is an estimate of the following integral:

$$\Phi(\alpha) = E \ln p(X|\alpha) = \int_{-\infty}^{\infty} p(X) \ln p(X|\alpha) dX \ .$$

Here,

$$p(X) = p(x_{t_0}, x_{t_1}, \ldots, x_{t_N})$$

is the α-independent, true joint density function of the process x_t at $t \in \Theta$, and the parameters of the model and intensities of the random deviations are incorporated into the vector α. An estimate of the unknown vector α is given by

$$\max_{\alpha} \Phi(\alpha) = \Phi(\alpha^*) \ . \tag{7.15}$$

Assume the existence of a vector $\alpha^* \in D$ such that for almost all $x, p(X|\alpha^*) = p(X)$ exists. Then we can choose the vector α from the maximum function condition given in equation (7.15). The validity of this equality follows from the work of Rao (1973). For any two densities $p(X)$ and $q(X)$,

$$\int_x p(x) \ln p(x) dx \geq \int_x p(x) \ln q(x) dx \ , \tag{7.16}$$

and equality takes place if, and only if, $p(x) = q(x)$ for almost all x. The inequality in equation (7.16) is the basis of the proof of the following theorem (Zuev, 1988).

Theorem 7.3
Let the sets D and Θ be such that for any α_1 and α_2, where $\alpha_1, \alpha_2 \in D \subset \mathbb{R}^l$, and $\alpha_1 \neq \alpha_2$, there always exists a term t_k, where $t_k \in \Theta$, such that $x_{t_k}(\alpha_1) \neq x_{t_k}(\alpha_2)$. Then, $\Phi(\alpha^*) > \Phi(\alpha)$ for all $\alpha \neq \alpha^*$, $\alpha \in D$.

It follows from our considerations that the following theorem is valid.

Theorem 7.4
The sequence of estimates $\{\alpha_m, \ m = 1, 2, \ldots\}$, as defined by the condition

$$\max_{\alpha \in D_1} \Phi_m = \Phi_m(\alpha_m)$$

(where D_1 is a parallelepiped) has, with unit probability, the limit point α^*. Here, D_1 is the set of parameter values and of perturbation intensities. The proof of this fact is based on theorem 7.3 and the continuity of $\Phi(\alpha)$.

7.4   Estimate Properties

Suppose that the conditions of theorem 7.3 are valid. Let us denote the derivative matrices as follows:

$$A(\alpha) = \left(\frac{\partial}{\partial \alpha^i}\Phi(\alpha)\right) \quad B(\alpha) = \left(\frac{\partial^2}{\partial \alpha^i \partial \alpha^j}\Phi(\alpha)\right)$$

$$C^r(\alpha) = \left(\frac{\partial^3}{\partial \alpha^r \partial \alpha^i \partial \alpha^j}\Phi(\alpha)\right) \quad i, j, r = 1, 2, \ldots, L \ . \tag{7.17}$$

The existence of these derivatives follows from the differentiation of the solution $x_t(\alpha)$ by parameters $(\partial/\partial \alpha)$. The same matrices for $\Phi_m(\alpha)$ are denoted by $A_m(\alpha)$, $B_m(\alpha)$, and $C_m^r(\alpha)$. Denote also $G(\alpha) = -E B_m(\alpha)$. It can be shown that

$$G(\alpha) = -E B_m(\alpha) = E A_m(\alpha) A_m^T(\alpha) \tag{7.18}$$

is valid.

Theorem 7.5
The estimate α_m of the unknown vector of parameters α^* is asymptotically efficient, asymptotically unbiased, and asymptotically normal.

Proof
First of all, we note that for any symmetric matrix C^r the following relationship is fulfilled (Lancaster, 1969):

$$\Delta \alpha^T C^r(\alpha) \Delta \alpha \leq \Lambda^r \Delta \alpha^T \Delta \alpha \ ,$$

where $\Delta\alpha$ is an arbitrary vector and Λ^r, given by $\Lambda^r = \max$ $(\lambda_1^r, \ldots, \lambda_n^r)$, is the maximal eigenvalue of matrix C^r. Let

$$\max_r \Lambda^r(\alpha) \leq H \ .$$

Then, for any r,

$$\Delta\alpha^T C^r(\alpha)\Delta\alpha < H\Delta\alpha^T\Delta\alpha \ .$$

Keeping this in mind, we expand the vector function $A_m(\alpha)$ in a Taylor series at the point α^*:

$$A_m(\alpha_m) = A_m(\alpha^*) + B_m(\alpha^*)\Delta\alpha_m + \frac{1}{2}Q(\alpha_m, \alpha^*)H_m\Delta\alpha_m^T\Delta\alpha_m$$

where $\Delta\alpha_m = \alpha_m - \alpha^*$ and $Q(\alpha_m, \alpha^*)$ is a row vector, the absolute values of elements of which are smaller than 1. Since α_m satisfies the condition

$$\max_\alpha \Phi_m(\alpha) = \Phi(\alpha_m) \ ,$$

then $A_m(\alpha_m) = 0$ means that

$$A_m(\alpha^*) + B_m(\alpha^*)(\alpha_m - \alpha^*) + \frac{1}{2}Q(\alpha_m, \alpha^*)H_m\Delta\alpha_m^T\Delta\alpha_m = 0 \ .$$

One can represent this relation in the following way:

$$-A_m(\alpha^*) = -G(\alpha^*)(\alpha_m - \alpha^*) + [B_m(\alpha^*) + G(\alpha^*)](\alpha_m - \alpha^*)$$
$$+ \frac{1}{2}Q(\alpha_m, \alpha^*)H_m(\alpha_m - \alpha^*)^T(\alpha_m - \alpha^*) \ .$$

From this, keeping in mind that $G(\alpha^*)$ is non-singular [see equation (7.17)], we have

$$(\alpha_m - \alpha^*) = G^{-1}(\alpha^*)A_m(\alpha^*) + G^{-1}(\alpha^*)[B_m(\alpha^*) + G(\alpha^*)]$$
$$\times(\alpha_m - \alpha^*) + \frac{1}{2}G^{-1}(\alpha^*)[Q(\alpha_m, \alpha^*)H_m$$
$$\times(\alpha_m - \alpha^*)^T(\alpha_m - \alpha^*)] \ .$$

In accordance with theorem 7.4, $\alpha_m \to \alpha^*$ with probability 1, and, therefore, the last term in this relationship tends to zero if $m \to \infty$. The second term on the right-hand side tends to zero. Also, according to equation (7.18), as $m \to \infty$, $B_m(\alpha)$ tends to $-G(\alpha^*)$ and $\alpha_m \to \alpha^*$. So, for sufficiently large m, $m > m^*$, one can write

$$(\alpha_m - \alpha^*) = G^{-1}(\alpha^*)A_m(\alpha^*) \ . \tag{7.19}$$

From this we find that

$$E(\alpha_m - \alpha^*) = G^{-1}(\alpha^*)EA_m(\alpha^*) \ .$$

However,

$$EA_m(\alpha^*) = \nabla_\alpha \Phi(\alpha^*) \ .$$

Taking into account that $\Phi(\alpha^*) > \Phi(\alpha)$ for any $\alpha \in D$, $\alpha \neq \alpha^*$, we have

$$E(\alpha_m - \alpha^*) = 0 \ .$$

This means that the maximum-likelihood estimate, α_m, of vector α^* is asymptotically unbiased. For covariance of the estimate,

$$E(\alpha_m - \alpha^*)(\alpha_m - \alpha^*)^T = EG^{-1}(\alpha^*)A_m(\alpha^*)A_m^T(\alpha^*)G^{-1}(\alpha^*) \ .$$

Keeping in mind equation (7.18), we find that

$$E(\alpha_m - \alpha^*)(\alpha_m - \alpha^*)^T = G^{-1}(\alpha^*) \ .$$

Expression (7.18) can be rewritten as

$$\begin{aligned}
G(\alpha^*) &= EA_m(\alpha^*)A_m^T(\alpha^*) = E\nabla_\alpha \Phi_m(\alpha^*)(\nabla_\alpha \Phi_m(\alpha^*))^T \\
&= E(\nabla_\alpha \ln p(X_m|\alpha))(\nabla_\alpha| \ln p(X_m|\alpha))^T \ .
\end{aligned}$$

In accordance with the Cramer–Rao inequality, $G^{-1}(\alpha^*)$ is the lower boundary for the matrix $E(\alpha_m - \alpha^*)(\alpha_m - \alpha^*)^T$. Therefore, the estimate α_m is asymptotically efficient.

Asymptotic normality of the estimate follows directly from equation (7.19) because, according to the Lindenberg–Levi theorem (Cramer, 1946), the asymptotic distribution $A_m(\alpha^*)$ is normal and the transformation (7.19) is linear. So the estimate of model parameters, α_m, is asymptotically unbiased, asymptotically efficient, and asymptotically normal with parameters

$$E\alpha_m = \alpha^*, \ \ \text{cov}(\alpha_m, \alpha_m) = G^{-1}(\alpha^*) \ .$$

7.5   Confidence Limits

According to the model in equation (7.8), $u_t = x_t^\varepsilon - x_t(\alpha)$ is a Gauss–Markov process. Based on this, we can derive confidence limits. First of all, the following theorem is valid.

Theorem 7.6

The sum of quadratic forms is given by

$$S(\alpha^*, \Gamma^*) = \sum_{i=1}^{N} \left(u_{t_i} - \mu_{t_i t_{i-1}}(\alpha^*)\right)^T K_{t_i t_{i-1}}^{-1}(\alpha^*, \Gamma^*)$$
$$\times \left(u_{t_i} - \mu_{t_i t_{i-1}}(\alpha^*)\right) ,$$

where $\mu_{st} \in \mathbb{R}^n$ has a χ^2_{Nn} distribution with the number of degrees of freedom being Nn.

Proof

We show first that the random values

$$u_{t_1} \quad - \quad E(u_{t_1}|u_{t_0})$$
$$u_{t_2} \quad - \quad E(u_{t_2}|u_{t_1})$$
$$\vdots$$
$$u_{t_N} \quad - \quad E(u_{t_N}|u_{t_{N-1}}) \tag{7.20}$$

are independent. This follows from the model in equation (7.8), the solution of which one can write in the form

$$u_t = R(t, s)u_s + \int_s^t R(t, \tau)dw_\tau, \quad t > s ,$$

where $R(t, s)$ is a fundamental matrix which satisfies the system of equations

$$\frac{d}{dt}R(t, s) = f_x(x_t(\alpha), \alpha)R(t, s), \quad R(s, s) = I .$$

Note that the conditional expectation $\mu_{\tau s} = E(u_\tau|u_s)$ satisfies the equation

$$\frac{d}{d\tau}\mu_{\tau s} = f_x(x_\tau(\alpha), \alpha)\mu_{\tau s}$$

where $\mu_{ss} = u_s$ and $\tau \geq s$. From this we can write

$$\mu_{ts} = R(t, s)u_s$$
$$u_t - R(t, s)u_s = u_t - E(u_t|u_s) = \int_s^t R(t, \tau)dw_\tau .$$

This means that

$$u_{t_i} - E(u_{t_i}|u_{t_{i-1}}) = \int_{t_{i-1}}^{t_i} R(t_i, \tau)dw_\tau .$$

Since w_t is a Gaussian process with independent increments, the independence of terms in equation (7.20) follows. However, the quadratic forms

$$(u_{t_i} - E(u_{t_i}|u_{t_{i-1}}))^T K_{t_i t_{i-1}}^{-1}(u_{t_i} - E(u_{t_i}|u_{t_{i-1}})), \quad i = 1, 2, \ldots, N$$

are also independent. Assume that $K_{t_i t_{i-1}}$ is not singular ($i = 1, 2, \ldots, N$). Let us consider one such quadratic form. The term

$$v_{t_i} = u_{t_i} - E(u_{t_i}|u_{t_{i-1}})$$

has a Gaussian distribution and zero mean. Let $v_{t_i} = A_i z$, where z is an n-dimensional, normally distributed, random vector such that $Ez = 0$ and cov $(z, z) = I$. Choose the matrix A_i such that

$$\mathrm{cov}(v_{t_i}, v_{t_i}) = E v_{t_i} v_{t_i}^T = E A_i z z^T A_i^T = A_i A_i^T = K_{t_i t_{i-1}} = K_i \ .$$

Then

$$\begin{aligned}
v_{t_i}^T K_i^{-1} v_{t_i} &= (A_i z)^T (A_i A_i^T)^{-1}(A_i z) \\
&= z^T A_i^T (A_i^T)^{-1} A_i^{-1} A_i z = z^T z = \sum_{i=1}^{N}(z^i)^2 \ ,
\end{aligned}$$

where z^i is the ith element of the vector z. Note that matrix A_i exists because K_i is a covariance matrix that is symmetrical and assumed positive definite (Rao, 1973). So the quadratic form can be represented as a sum of n quadratics of normally distributed, random variables with zero mean and unit covariance. It follows from this that the sum of such quadratic forms, $S(\alpha^*, \Gamma^*)$, can be represented as the sum of Nn quadratics of the random terms in question. Therefore, the sum has a χ^2_{Nn} distribution with Nn degrees of freedom.

Let $\alpha = \alpha^*$ and the theoretical distribution of u_t correspond to the real one. Using sampled trajectories,

$$X_m = \{X^i, i = 1, 2, \ldots, m\}, \quad X^i = \{x_t^i, t \in \Theta\} \ ,$$

construct the following sum:

$$S(\alpha^*, \Gamma^*, X_m) = \frac{1}{L} \sum_{i=1}^{m} \sum_{j=1}^{N}(u_j^i - \mu_j^i(\alpha^*))^T K_j^{-1}(\alpha^*, \Gamma^*)(u_j^i - \mu_j^i(\alpha^*))$$

where

$$u_j^i = x_{t_j}^i - x_{t_j}(\alpha^*)$$
$$\mu_j^i(\alpha^*) = E(x_{t_j} - x_{t_j}(\alpha^*)|x_{j-1}^i - x_{t_{j-1}}(\alpha^*))$$
$$K_j(\alpha^*, \Gamma^*) = K_{t_j t_{j-1}}(\alpha^*, \Gamma^*)$$

and the moments μ and K are given by equation (7.13). Since the trajectories from X_m are independent, it follows from theorem 7.5 that the term $LS(\alpha^*, \Gamma^*, X_m)$ has the distribution χ_r^2 with $r = mnN$ degrees of freedom.

Let χ_p^2 satisfy the condition

$$P(\chi_r^2 > \chi_p^2) = p \ ,$$

where p is a small probability (say, $p = 0.05$). Then, the fact that the event $LS(\alpha^*, \Gamma^*, X_m) > \chi_p^2$ is unlikely is explained by purely random factors. Therefore, if

$$S(\alpha^*, \Gamma^*, X_m) > \frac{\chi_p^2}{r} \ ,$$

it is unlikely that the model corresponds to real data. On the other hand, the inequality

$$S(\alpha, \Gamma, X_m) \le \frac{\chi_p^2}{r}$$

is valid at values of α and Γ which form some set $Q(X_m)$, all points of which with probability $(1-p)$ can be considered as permissible parameter values . In other words, this inequality defines the $(1-p)100\%$ confidence set for the parameters. This result is the basis for the construction of confidence intervals and criteria of fitness between model and data.

7.6 Filtering Problem

In the case of the estimation of α^* by X_m with measurement error, we will use filtering theory (Balakrishnan, 1984; Liptser and Shiryaev, 1977). Recall that for the description of the deviations of the real process x_t^ε from the solution of the deterministic model $x_t(\alpha^*)$ we can use the system of stochastic differential equations:

$$\frac{d}{dt}u_t = f_x(x_t(\alpha^*), \alpha^*)u_t + \frac{d}{dt}w_t \ , \tag{7.21}$$

where $u_o = 0, t \in [0, T]$, and

$$u_t = x_t^\varepsilon - x_t(\alpha^*)$$
$$\frac{d}{dt} x_t(\alpha^*) = f_x(x_t(\alpha^*), \alpha^*)$$

for $x_0 = c$ and $\alpha^* \in \mathbb{R}^L$.

Let X_m be the set of realizations defined on Θ for the following stochastic process:

$$x_t = x_t^\varepsilon + \eta_t \tag{7.22}$$

where $t \in \Theta, \eta_t \in \mathbb{R}^n, \eta_t$ is Gaussian white noise for discrete time, $E\eta_t = 0$, and $E\eta_t \eta_t^T = \Gamma_0$ for all $t \geq 0$. Now, for the model description of the process u_t we can represent equation (7.22) in the form

$$x_t = u_t + x_t(\alpha^*) + \eta_t$$

where $t \in \Theta$.

Let us now take into account the fact that trajectories are defined on the set Θ. In this connection we go from the continuous model in equation (7.21) to the discrete model. For this we can use the fundamental matrix $R(t, s)$ of the system of linear equations, and write at $t > s \geq 0$ the solution of the form

$$u_t = R(t, s)u_s + \int_s^t R(t, v) dw_v$$
$$\frac{d}{dt} R(t, s) = f_x(x_t(\alpha^*), \alpha^*) R(t, s), \quad R(s, s) = I \ .$$

Denote

$$u_i = u_{t_i}$$
$$W_t = \int_{t_{i-1}}^{t_i} R(t, v) dw_v$$

where, for all i and j, $t_i \in \Theta, i = 1, 2, \ldots, N$; $\text{cov}(W_i, W_j) = 0, i \neq j$; $\text{cov}\,(u_i, W_i) = 0$, $\text{cov}\,(W_i, \eta_j) = 0$, and $\text{cov}\,(\eta_i, u_j) = 0$. So we come to the following model (Balakrishnan, 1986):

$$u_i = R(t_i, t_{i-1})u_{i-1} + W_i, \quad u_0 = 0$$
$$x_i = u_i + \eta_i + x_i(\alpha^*), \quad i = 1, 2, \ldots \ . \tag{7.23}$$

In terms of filtering theory, u_t is the signal (unobserved variable) and the first equation is the signal model. The observed variable is x_t: it represents the sum of the signal u_t and error η_t. The problem is to define the signal parameters as values of the observed variable $\{x_i, i = 1, 2, \ldots, N\}$. Consider the maximum-likelihood ratio (Balakrishnan, 1984):

$$L(X|\alpha) = \frac{P(X|H_1, \alpha)}{P(X|H_0)} \rightarrow \max_{\alpha \in D} \; ,$$

where $P(X|H_1, \alpha)$ is the conditional joint density of $X = \{x_i, i = 1, 2, \ldots, N\}$ with $x_i = u_i + \eta_i + x_i(\alpha)$ (hypothesis 1), and $p(X|H_0)$ is the same conditional density that corresponds to hypothesis H_0. Under hypothesis H_0 the signal is absent in the observed values x, that is, $x_i = \eta_i$. We can write

$$p(X|H_0) = \prod_{i=1}^{N} P(x_i) \; , \tag{7.24}$$

where $p(x_i)$ is a Gaussian density with zero mathematical expectation and covariance matrix Γ_0. The conditional density function $p(X|H_1, \alpha)$ can be found using the fact that random values

$$
\begin{aligned}
z_1 &= x_1 \\
z_2 &= x_2 - E(x_2|x_1) \\
z_3 &= x_3 - E(x_3|x_2, x_1) \\
&\;\;\vdots \\
z_N &= x_N - E(x_N|x_{N-1}, \ldots, x_1)
\end{aligned}
$$

are independent. This follows from the equations in (7.24) which are Gram-Schmidt orthogonalized processes (Rao, 1973). Due to this

$$p(X|H_1, \alpha) = \prod_{i=1}^{N} p(z_i) \; ,$$

where $p(z_i)$ is a Gaussian density with zero mean and covariance matrix G_i that is needed to define the process. First, the conditional expectation is

$$
\begin{aligned}
E(x_i|x_{i-1}, x_{i-2}, \ldots, x_1) &= E(u_i + \eta_i + x_i(\alpha^*)|x_{i-1}, x_{i-2}, \ldots, x_1) \\
&= x_i(\alpha^*) + E(u_i|x_{i-1}, x_{i-2}, \ldots, x_1) \; .
\end{aligned}
$$

Using u_i and the recurrent formulae in equation (7.23), we have

$$E(x_i|x_{i-1}, x_{i-2}, \ldots, x_1)$$
$$= x_i(\alpha^*) + E(R(t_i, t_{i-1})u_{i-1} + w_i|x_{i-1}, x_{i-2}, \ldots, x_i)$$
$$= x_i(\alpha^*) + R(t_i, t_{i-1})E(u_{i-1}|x_{i-1}, x_{i-2}, \ldots, x_i)$$
$$= x_i(\alpha^*) + R(t_i, t_{i-1})\hat{u}_{i-1} \ ,$$

where

$$\hat{u}_i = E(u_i|x_i, \ldots, x_1)$$

is a signal estimate based on observations $x_1, x_2, \ldots, x_i$, that is, the solution of the filtering problem. Recurrent relationships exist for $\hat{u}_i$ known as the recursive Kalman filter:

$$\hat{u}_i = R(t_i, t_{i-1})\hat{u}_{i-1} + P_i\Gamma_0^{-1}(x_i - R(t_i, t_{i-1})\hat{u}_{i-1} - x_i(\alpha^*)) \qquad (7.25)$$

where $\hat{u}_0 = 0$,

$$\begin{aligned} P_i &= (I + H_{i-1}\Gamma_0^{-1})^{-1}H_{i-1} \\ H_k &= R(t_k, t_{k-1})P_kR^T(t_k, t_{k-1}) + Q(t_k, t_{k-1}) \\ Q(t_k, t_{k-1}) &= Ew_kw_k^T \ . \end{aligned} \qquad (7.26)$$

So, for z_i we have

$$z_i = x_i - x_i(\alpha^*) - R(t_k, t_{i-1})\hat{u}_{i-1} \ ,$$

where the vector $\hat{u}_{i-1}$ is defined by equation (7.25). In order to find $\mathrm{cov}(z_i, z_i)$, we include x_i from equation (7.23) in this expression so that

$$\begin{aligned} V_i &= \mathrm{cov}(z_i, z_i) \\ &= \mathrm{cov}(\eta_i + u_i - R(t_i, t_{i-1})\hat{u}_{i-1}, \eta_i + u_i - R(t_i, t_{i-1})\hat{u}_{i-1}) \\ &= \mathrm{cov}(\eta_i, \eta_i) + \mathrm{cov}(u_i - R(t_i, t_{i-1})\hat{u}_{i-1}, u_i - R(t_i, t_{i-1})\hat{u}_{i-1}) \\ &= \Gamma_0 + H_i \ , \end{aligned}$$

where H_i is the covariance matrix of the predicted error u_i based on previous observation. For the likelihood relationship $L(X|\alpha)$, we find that

$$\begin{aligned} L(X|\alpha) &= \exp\left[-\frac{1}{2}\left(\sum_{i=1}^N [(x_i - x_i(\alpha) - R(t_i, t_{i-1})\hat{u}_{i-1})^T\right.\right. \\ &\quad \times V_i^{-1}(x_i - x_i(\alpha) - R(t_i, t_{i-1})\hat{u}_{i-1})] \\ &\quad \left.\left. - \sum_{i=1}^N x_i^T\Gamma_0 x_i + \sum_{i=1}^N \ln\det(V_i) - \sum_{i=1}^N \ln\det(\Gamma_0)\right)\right] \ . \end{aligned}$$

This expression represents the likelihood relationship derived for the solution of the parameter estimation problem using one trajectory. For $L(X_m|\alpha)$ we have

$$l(X_m|\alpha) = \prod_{i=1}^{m} L(X^i|\alpha) \ .$$

Example

Consider the model

$$\frac{d}{dt}x_t(\lambda) = -\lambda x_t(\lambda) \ ,$$

where $x_0 = 1$ and $t \geq 0$. For a description of the deviations $u_t = x_t^\varepsilon - x_t(\lambda)$, we obtain the following stochastic equation:

$$\frac{d}{dt}u_t = -\lambda u_t + Bx_t(\lambda)dw_t \ ,$$

where $u_0 = 0$. Representation of equation (7.23) in this case has the form

$$u_t = R(t,s)u_s + B\int_s^t R(t,v)x_v dw_v \ ,$$

where $R(t,s)$ satisfies the equation

$$\frac{d}{dt}R(t,s) = -\lambda R(t,s), \quad R(s,s) = 1 \ .$$

Taking into account that $R(t,\tau) = R(t,s)R^{-1}(\tau,s)$, we derive

$$u_t = u_s R(t,s) + BR(t,s)\int_s^t R^{-1}(v,s)x_v dw_v \ .$$

Since $R(t,s) = e^{-\lambda(t-s)}$, we conclude that

$$\begin{aligned} u_t &= u_s e^{-\lambda(t-s)} + Be^{-\lambda t}\int_s^t dw_v = u_s e^{-\lambda(t-s)} \\ &\quad + Be^{-\lambda t}(w_t - w_s) \ . \end{aligned}$$

From this result, we derive the following expressions for the model of signal formation and measurement processes:

$$\begin{aligned} u_i &= R(t_i, t_{i-1})u_{i-1} + W_i \\ x_i &= u_i + \eta_i + e^{-\lambda t_i} \ , \end{aligned}$$

where

$$Eη_i^2 = γ$$
$$R(t_i, t_{i-1}) = \exp[-λ(t_i - t_{i-1})]$$
$$w_i = B(w_{t_i} - w_{t_{i-1}})\exp(-λt_i) \ .$$

In equation (7.26)

$$Q(t_k, t_{k-1}) = EW_k^2 = B^2(t_k - t_{k-1})\exp(-2λt_k) \ .$$

In the next chapter we consider the iterative algorithms for the estimation of parameters. In fact, the method of parameter estimation considered above is based on maximization of the likelihood function that implicitly depends on the values estimated. This complicates the solution of the extremum problem and leads to large computational difficulties.

Chapter 8

Marchuk–Zuev Identification Problem

In Chapter 7 the problem of statistical estimation of model parameters is discussed. It is a difficult problem, because the likelihood function depends on the parameters of the model in the implicit form. Therefore, in this chapter an effective numerical algorithm which uses the adjoint equations is constructed. A deterministic case was studied by Marchuk (1981) and a stochastic one by Zuev (1986, 1988).

8.1 The Deterministic Case

Let the model be represented by

$$\frac{d}{dt}x_t = f(x_t, \alpha), \quad x_0 = c, \quad t \in [0, T]$$

$$x_t \in \mathbb{R}^n, \quad \alpha \in \mathbb{R}^L . \tag{8.1}$$

The vector-function $f(x, \alpha)$ has bounded first and second partial derivatives with respect to x and α. In contrast to the previous discussions, we suppose that the observed trajectory of state variables

$$X = \{x_t, t \in \theta\}, \quad \theta = \{t_0, t_1, \ldots, t_N\}$$

belongs to the set of solutions of the model in equation (8.1). In other words, we suppose that there exists a vector $\alpha^* \in D \subset \mathbb{R}^L$ such that

$$x_t(\alpha^*) = x_t, \quad t \in [0, T] .$$

Assume that $\alpha^* = \alpha_0 + \delta\alpha$, where α_0 is a known vector and $\delta\alpha$ is small in comparison with α_0. This means that α_0 is a "good" initial approximation for the problem solution, and we want to define the vector $\delta\alpha$ using X.

Assuming $\delta\alpha$ is small, we can then introduce a small parameter $\varepsilon > 0$ representing

$$\delta\alpha = \varepsilon\Delta\alpha, \quad \Delta\alpha \in \mathbb{R}^L .$$

As a result, the model in equation (8.1) will have the form

$$\frac{d}{dt}x_t(\alpha_0 + \varepsilon\Delta\alpha) = f[x_t(\alpha_0 + \varepsilon\Delta\alpha), \ \alpha_0 + \varepsilon\Delta\alpha]$$
$$x_0 = c . \tag{8.2}$$

Assuming that the relationship is smooth, we can write the solution $x_t(\alpha^*)$ in the form of an expansion according to degrees of ε:

$$x_t(\alpha^*) = x_t^{(0)} + \varepsilon x_t^{(1)} + \varepsilon^2 x_t^{(2)} + \dots . \tag{8.3}$$

Substitute equation (8.3) into equation (8.2), expand the right-hand side in powers of the small parameter ε, where $\varepsilon > 0$, up to terms of the order of $O(\varepsilon^N)$,

$$\frac{d}{dt}x_t^{(0)} + \varepsilon\frac{d}{dt}x_t^{(1)} + \dots$$
$$= f(x_t^{(0)}, \alpha_0) + \varepsilon f_x(x_t^{(0)}, \alpha_0)x_t^{(1)} + \varepsilon f_\alpha(x_t^{(0)}, \alpha_0)\Delta\alpha + \dots ,$$

and equate the terms with the same powers of ε, where $\varepsilon > 0$, to obtain the system

$$\frac{d}{dt}x_t^{(0)} = f(x_t^{(0)}, \alpha_0), \ x_0^{(0)} = c$$
$$\frac{d}{dt}x_t^{(1)} = f_x(x_t^{(0)}, \alpha_0)x_t^{(1)} + f_\alpha(x_t^{(0)}, \alpha_0)\Delta\alpha, \ x_0^{(1)} = 0 ,$$

and so on.

Neglecting terms of the order of ε^2 and higher, for

$$\delta x_t \approx x_t(\alpha^*) - x_t(\alpha_0)$$
$$\equiv x_t(\alpha_0 + \delta\alpha) - x_t(\alpha_0) ,$$

we have

$$\frac{d}{dt}\delta x_t = f_x(x_t^0, \alpha_0)\delta x_t + f_\alpha(x_t^0, \alpha_0)\delta\alpha \tag{8.4}$$

where $t \in [0,T]$, $\delta x_0 = 0$, and $x_t^0 = x_t(\alpha_0)$.

For the system given in equation (8.4) we can write the adjoint equation

$$\frac{d}{dt} y_t = -f_x^T(x_t^0, \alpha_0)y_t + Q(t)$$

$$t \in [0,T], \quad y_T = 0 \ . \tag{8.5}$$

The function $Q(t)$ will be defined below. Multiplying equation (8.4) by y_t and equation (8.5) by δx_t, and integrating from 0 to T, we have

$$\langle f_x(x_t^0, \alpha_0)\delta x_t, y_t \rangle - \langle f_x^T(x_t^0, \alpha_0)y_t, \delta x_t \rangle = 0 \ ,$$

where $\langle . \rangle$ is an inner product. Then, with $y_T = 0$ and $\delta x_0 = 0$, we derive

$$\int_0^T \left(\left\langle y_t, \frac{d}{dt}\delta x_t \right\rangle + \left\langle \delta x_t, \frac{d}{dt}y_t \right\rangle \right) dt = \int_0^T \frac{d}{dt}\langle y_t, \delta x_t \rangle = \langle y_t, \delta x_t \rangle \Big|_0^T$$

$$= \int_0^T \langle f_\alpha(x_t^0, \alpha_0)\delta\alpha, y_t \rangle dt + \int_0^T \langle Q(t), \delta x_t \rangle dt \ .$$

Finally,

$$\int_0^T \langle f_\alpha(x_t^0, \alpha_0)\delta\alpha, y_t \rangle + \int_0^T \langle Q(t), \delta x_t \rangle dt = 0 \ . \tag{8.6}$$

Expression (8.6) holds for linear approximation and represents the explicit relationship between known deviations

$$\delta x_t = x_t - x_t(\alpha_0)$$

and an unknown vector $\delta\alpha$. Indeed, let us choose a vector function $Q(t)$ such that all its components, except for $q^i(t)$, are equal to zero, and $q^i(t)$ is a delta function: that is,

$$Q(t) = Q^i(t,s) = (0,0,\ldots,q^i(t,s),0,0,\ldots,0)^T$$

$$q^i(t,s) = \delta(t-s) \ .$$

In this case, the equality (8.6) acquires the following form:

$$\left\langle \int_0^T f_\alpha^T(x_T^0, \alpha_0)y_{ist}dt, \delta\alpha \right\rangle + \delta x_s^i = 0 \ , \tag{8.7}$$

where

$$\delta x_s^i = x_s^i - x_s^i(\alpha_0)$$

and y_{ist} satisfies equation (8.5) in which the function $Q(t)$ is defined as above.

Equality (8.7) represents the linear algebraic equation with respect to the vector $\delta\alpha$. In order to have a system of such equations for the computation of $\delta\alpha$, expression (8.7) should be written for known deviations of the model, that is, for $s = t_0, t_1, \ldots, t_N (i = 1, 2, \ldots, n)$, and N should equal 1 (the number of unknown parameters). Let $\delta\alpha_0$ be the solution of this system. We represent the unknown vector α^* as $\alpha^* = \alpha_0 + \delta\alpha$. As a result of the calculation of $\delta\alpha$ based on linearization, we obtain a value of $\delta\alpha_0$ which is different from $\delta\alpha$. Then, replace α^* by the vector

$$\alpha_1 = \alpha_0 + \delta\alpha_0 \ .$$

Therefore, α_1 is a first approximation of the unknown vector α^*. Now we can write an iterative process for the α^* calculation:

$$\alpha_{n+1} = \alpha_n + \delta\alpha_n \ .$$

It is clear that for sufficiently small values of $|\,\alpha^* - \alpha_0\,|$, the linearization executed will be valid and the iterative process will converge: that is,

$$\lim_{n \to \infty} \alpha_n = \alpha^* \ .$$

However, this question deserves separate consideration.

8.2   Convergence: The Scalar Case

Suppose that $n = 1$ and $N > 1$. It will become clear on further consideration that all these speculations are valid for $n > 1$. Let $F(\alpha)$ be a vector function, the components of which are the deviations

$$F^i(\alpha) = x_{t_i} - x_{t_i}(\alpha), \quad i = 1, 2, \ldots, N \ .$$

As long as

$$\{x_t, t \in [0, T]\}$$

belongs to the set of solutions of equation (8.1), then there exists an unknown vector $\alpha^* \in D$ such that $F(\alpha^*) = 0$. So, the system of nonlinear, algebraic equations $F(\alpha^*) = 0$ should be solved for the calculation of α^*.

Consider the iterative process. Let α_n be some approximation of α^*. Since $x_t(\alpha)$ is differentiable by α, there exists a matrix of derivatives

$$F'(\alpha) = \partial F^i(\alpha)/\partial \alpha^j \ ,$$

and we can write

$$| F(\alpha^*) - F(\alpha_n) - F'(\alpha_n)(\alpha^* - \alpha_n) | = O(| \alpha^* - \alpha_n |) .$$

If the term $| \alpha^* - \alpha_n |$ is small, we can write

$$F(\alpha_n) + F'(\alpha_n)(\alpha^* - \alpha_n) \approx F(\alpha^*) = 0 .$$

It is natural to choose the next approximation α_{n+1} as a solution of the equation

$$F(\alpha_n) + F'(\alpha_n)(\alpha_{n+1} - \alpha_n) = 0 ,$$

that is,

$$\alpha_{n+1} = \alpha_n - [F'(\alpha_n)]^{-1} F(\alpha_n) . \tag{8.8}$$

This is the well-known iterative process called Newton's method. The following theorem gives the convergence condition of this process and the accuracy of the calculation of α^*.

Theorem 8.1
Let

$$D_a = \{\alpha :| \alpha - \alpha^* |< a\} .$$

For some values of $a, a_1,$ and a_2, where $0 < a < a_1, a_2 < \infty$, the following conditions are valid:

$$\|F'(\alpha)^{-1}\| \le a_1$$

for $\alpha \in D_a \subset \mathbf{R}^L$ and

$$| F(u_1) - F(u_2) - F'(u_2)(u_1 - u_2) | \le a_2 | u_2 - u_1 |$$

for all $u_1, u_2 \in D_a$. Let $c = a_1 a_2$ and $b = \min(a, c^{-1})$. Then, for $\alpha_0 \in D_b$, the iterative process in equation (8.8) converges with the convergence rate

$$| \alpha_n - \alpha^* | \le c^{-1}(c | \alpha_0 - \alpha^* |)^{2^n} . \tag{8.9}$$

Let us now consider the iterative process in equation (8.8) and prove the following theorem.

Theorem 8.2

Assume that first and second partial derivatives of $f(x, \alpha)$ exist. Let the ith, jth element of a matrix R be given by

$$R^{ij}(\alpha) = \frac{\partial x_{t_i}}{\partial \alpha^j}, \quad i, j = 1, 2, \dots, L \ .$$

For all values of α, where $\alpha \in D_a$, the inequality

$$\|[R(\alpha)]^{-1}\| \leq a_1$$

is valid, and the initial point α_0 belongs to D_b, where D_a and D_b are defined by the conditions of theorem 8.1. Then the iteration process in equation (8.8) converges and the estimate given in equation (8.9) exists.

Proof

Denote the derivative of the solution of equation (8.1) with respect to the parameter α^i as x_{it}:

$$x_{it} = \partial x_t(\alpha)/\partial \alpha^i \ .$$

From this we derive equations for $x_{it}(i = 1, 2, \dots, L)$:

$$\frac{d}{dt} x_{it} = f_x(x_t, \alpha) x_{it} + f_{\alpha^i}(x_t, \alpha), \quad x_{i0} = 0$$

$$f_{\alpha^i}(x, \alpha) = \partial f(x, \alpha)/\partial \alpha^i \ . \tag{8.10}$$

Consider the adjoint equation

$$\frac{d}{dt} y_{st} = -f_x(x_t, \alpha) y_{st} + \delta(t - s)$$

$$t \in [0, T], \quad y_{ST} = 0 \ . \tag{8.11}$$

Multiply equation (8.10) by y_{st} and equation (8.11) by x_{it}, add the results and integrate on $[0, T]$. After transformations we have

$$x_{is} = -\int_0^T f_{\alpha^i}(x_t, \alpha) y_{st}(\alpha) dt \ .$$

To calculate $\delta\alpha_n$ by the iterative process we have the following system of equations:

$$\sum_{i=1}^l \delta\alpha_n^i \int_0^T f_{\alpha^i}(x_t(\alpha_n), \alpha_n) y_{st}(\alpha_n) dt + x_s - x_s(\alpha_n) = 0$$

where $s = t_1, t_2, \ldots, t_N$. From here we have

$$\sum_{i=1}^{l} \delta \alpha^i x_{is} = x - x_s(\alpha_n) \ .$$

The terms $x_{is}(i = 1, 2, \ldots, l; s = t_1, t_2, \ldots, t_N)$ form the matrix $R(\alpha_n)$, and the differences, $x_s - x_s(\alpha_n)$, form the vector $F(\alpha_n)$. Then it is possible to represent the last displayed expression as

$$R(\alpha_n)\delta \alpha_n = F(\alpha_n) \ .$$

As long as $R(\alpha_n)$ is non-singular, then

$$\delta \alpha_n = [R(\alpha_n)]^{-1} F(\alpha_n) \ .$$

With these notations, the iterative process in equation (8.8) takes the form

$$\alpha_{n+1} = \alpha_n + \delta \alpha_n = \alpha_n + R^{-1}(\alpha_n) F(\alpha_n) \ .$$

Recalling that $R(\alpha_n) = -F'(\alpha_n)$, we see from here that equation (8.9) is a Newtonian process. As long as we assume the existence of the second derivatives of the right-hand side of $f(x, \alpha)$, then for all values of t (where $t \geq 0$), $\alpha_1, \alpha_2 \in D_a \subset \mathbb{R}^L$, and we have

$$x_t(\alpha_1) = x_t(\alpha_2) + \sum_{i=1}^{L} x_{it}(\alpha_2)(\alpha_2^i - \alpha_1^i) + O(|\alpha_1 - \alpha_2|) \ .$$

From here it follows that the second condition of theorem 8.1 is satisfied. Therefore, the process in equation (8.8) converges and the estimate in equation (8.9) is valid. Similarly, the multidimensional case $n > L$ can be considered.

8.3 Statistical Estimation of Parameters

Now we use the same idea as used for the construction of an iterative process to calculate the estimate α^* in the case when the observed values of the model variables have random character. Moreover, we take into account those features of the data of immunological experiments which cause the measured values of state variables to be independent: that is, the matrix $\text{cov}(x_t, x_t)$ is diagonal and, for $s < t$,

$$P(x_t \mid x_s) = P(x_t)$$

where $s, t \in \theta$. This corresponds to the conditions in real experiments, while the values of separate state variables are derived by using measurements for different animals.

Then, let α_0 be some initial approximation with $\alpha^* = \alpha_0 + \delta\alpha$, where $\delta\alpha$ is small in comparison with α_0. Now, the stochastic model for the deviations u_t, given by $u_t = x_t - x_t(\alpha_0)$, has the form

$$\frac{d}{dt}u_t = f_x(x_t^0, \alpha_0)u_t + f_\alpha(x_t^0, \alpha_0)\delta\alpha + f_\alpha(x_t^0, \alpha_0)\frac{d}{dt}w_t$$

$$\frac{d}{dt}x_t^0 = f(x_t^0, \alpha_0), \quad x_0 = c, \quad u_0 = 0, \quad t \in [0, T] \ . \tag{8.12}$$

Here we take into account the fact that

$$f(x, \alpha) = F(x)\alpha = f_\alpha(x, \alpha)\alpha \ .$$

Also, $\text{cov}(w_t, w_t) = \Gamma t$. Following the methods mentioned above in the previous section, consider the adjoint system

$$\frac{d}{dt}y_{ist} = -f_x^T(x_t^0, \alpha_0)y_{ist} + Q^i(t, s)$$

$$i = 1, 2, \ldots, n, \quad s \in \theta, \quad t \in [0, T], \quad y_{isT} = 0 \ . \tag{8.13}$$

Taking the inner product of equation (8.12) and multiplying by y_{ist} and the inner product of equation (8.13) and multiplying by u_t, adding the results, and integrating the sum on the interval $[0, T]$, we obtain

$$\int_0^T \langle f_\alpha^T(x_t^0, \alpha_0)y_{ist}, \delta\alpha \rangle dt + u_s^i$$

$$= -\int_0^T \langle f^T(x_t^0, \alpha_0)y_{ist}, dw_t \rangle \ .$$

The properties of the random values

$$\int_0^T \langle f_\alpha^T(x_t^0, \alpha_0)y_{ist}, \delta\alpha \rangle dt + u_s^i, \quad s \in \theta$$

are defined by the properties of the stochastic integral

$$\int_0^T \langle f_\alpha^T(x_t^0, \alpha_0)y_{ist}, dw_t \rangle \ .$$

For simplicity, let the matrix Γ be diagonal. Then,

$$E(x_s^i - x_s^i(\alpha_0)) = \langle a_s^i(\alpha_0), \delta\alpha \rangle$$

$$\text{var}(x_s^i - x_s^i(\alpha_0)) = \langle b_s^i(\alpha_0), \Gamma \rangle, \quad s \in \theta \ ,$$

where

$$a_s^i(\alpha_0) = \int_0^T f_\alpha^T(x_t^0, \alpha_0) y_{ist} dt$$

$$\Gamma = (\gamma^{11}, \gamma^{22}, \ldots, \gamma^{LL})^T$$

and the vector $b_s^i(\alpha)$ consists of the following elements:

$$b_s^{ik}(\alpha_0) = \int_0^T \langle f_\alpha^k(x_t^0, \alpha_0) y_{ist} \rangle^2 dt \; .$$

Here, $f_\alpha^k(x, \alpha)$ is the kth row of the matrix $f_\alpha(x, \alpha)$. Again, for small deviations these values are nearly Gaussian, so for the derivation of asymptotically effective estimates for $\delta\alpha$ and Γ we can use the maximum-likelihood principle.

Analogous to Chapter 7, we find that the estimates of the unknown parameters are the coordinates of the following function minimum:

$$\Phi_m(\delta\alpha, \Gamma \mid \alpha_0) = \sum_{t \in \theta} \frac{1}{m} \sum_{j=1}^{m} \sum_{i=1}^{n} \left(\ln\langle b_t^i(\alpha_0), \Gamma \rangle \right.$$
$$\left. + \frac{[x_t^{ij} - x_t^i(\alpha_0) + \langle a_t^i(\alpha_0), \delta\alpha \rangle]^2}{\langle b_t^i(\alpha_0), \Gamma \rangle} \right) \; .$$

As a result, we have the likelihood function that explicitly depends on unknown parameters. The search of its extremum is not a complex computational problem. The iterative process considered in the previous section is now written as

$$\min_{\delta\alpha, \Gamma} \Phi_m(\delta\alpha, \Gamma \mid \alpha_k) = \Phi_m(\delta\alpha_k \Gamma_k \mid \alpha_k)$$

$$\alpha_{k+1} = \alpha_k + \delta\alpha_k, \quad k = 0, 1, 2, \ldots \; . \tag{8.14}$$

Using the same algorithm we can modify equation (8.14) to calculate an unknown initial value given by $x_0 = c$ (Zuev, 1988).

Now we discuss the iterative process convergence. It is easy to see that equation (8.14) is a simple iteration method:

$$\alpha_{k+1} = v(\alpha_k), \quad k = 0, 1, 2, \ldots$$

$$v(\alpha) = \alpha + \delta\alpha(\alpha) \; , \tag{8.15}$$

and the function $\delta\alpha(\alpha)$ satisfies the condition

$$\min_{\delta\alpha, \Gamma} \Phi_m(\delta\alpha, \Gamma \mid \alpha) = \Phi_m(\delta\alpha(\alpha), \Gamma(\alpha) \mid \alpha) \; .$$

We assume that the solution is unique: that is, $\delta\alpha$ and Γ belong to convex sets, and $\Phi_m(\delta\alpha, \Gamma \mid \alpha)$ is a strictly convex function.

The following result is well known in computational mathematics (Bakhvalov *et al.*, 1987).

Theorem 8.3

Let, for all $\alpha, \beta \in D$ and $\rho < 1$, the inequality

$$\|v(\alpha) - v(\beta)\| \le \rho\|\alpha - \beta\|$$

be fulfilled. Then, the iterative process of equation (8.15) converges to the solution $\overline{\alpha} = v(\overline{\alpha})$ and the following estimate is valid:

$$\|\alpha_k - \overline{\alpha}\| \le \frac{\rho^k}{1 - \rho} \|\alpha_0 - \overline{\alpha}\| \ . \tag{8.16}$$

Let us try to find the conditions for which inequality (8.16) is valid. Let α_1 and α_2 belong to a small neighborhood around $\overline{\alpha}$: that is,

$$\begin{aligned}
\alpha_1 &= \overline{\alpha}(1 - \lambda) + \lambda\beta_1 \\
\alpha_2 &= \overline{\alpha}(1 - \lambda) + \lambda\beta_2
\end{aligned}$$

where $\beta_1, \beta_2 \in \mathbb{R}^l$, $0 < \lambda \ll 1$, and $\beta_1 \ne \beta_2$. Then, assuming λ to be close to zero, we can write, for $\beta \ne \overline{\alpha}$,

$$\begin{aligned}
v(\overline{\alpha}(1 - \lambda) + \lambda\beta) &= v(\overline{\alpha}) + v_\alpha(\overline{\alpha})(\beta - \overline{\alpha})\lambda \\
&= \overline{\alpha} + \lambda v_\alpha(\overline{\alpha})(\beta - \overline{\alpha}) \ ,
\end{aligned}$$

where $v_\alpha(\overline{\alpha})$ is a matrix of derivatives of the vector function $v(\alpha)$ at $\alpha = \overline{\alpha}$. Based on this we obtain

$$\begin{aligned}
\|v(\alpha_1) - v(\alpha_2)\| &= \|v_\alpha(\overline{\alpha})(\alpha_1 - \alpha_2)\| \\
&\le \|v_\alpha(\overline{\alpha})\| \cdot \|\alpha_1 - \alpha_2\| \ .
\end{aligned}$$

Clearly, the inequality

$$\|v_\alpha(\overline{\alpha})\| < 1 \tag{8.17}$$

is the convergence condition. It is simple to show that

$$v_\alpha(\overline{\alpha}) = I - A^{-1}(\overline{\alpha})\partial G_\alpha(\overline{\alpha}) \ ,$$

where matrix $A(\overline{\alpha})$ has the elements

$$
\begin{aligned}
A^{rk}(\overline{\alpha}) &= \sum_{t\in\theta} \frac{1}{m} \sum_{j=1}^{m} \sum_{i=1}^{n} \frac{a_t^{ik}(\overline{\alpha})a_t^{ir}(\overline{\alpha})}{\langle b_t^i(\overline{\alpha}), \overline{\Gamma}\rangle} \\
&= \sum_{t\in\theta} \sum_{i=1}^{n} \frac{a^{ik}(\overline{\alpha})a^{ir}(\overline{\alpha})}{\langle b_t^i(\overline{\alpha})\overline{\Gamma}\rangle}
\end{aligned}
\tag{8.18}
$$

and vector $G(\overline{\alpha})$ is given by

$$
G^k(\overline{\alpha}) = \sum_{t\in\theta} \frac{1}{m} \sum_{j=1}^{m} \sum_{i=1}^{n} \frac{x_t^{ij} - x_t^i(\overline{\alpha})}{\langle b_t^i(\overline{\alpha}), \overline{\Gamma}\rangle} \, a_t^{ik}(\overline{\alpha}) \ .
\tag{8.19}
$$

If we differentiate expression (8.19), we find that the matrix $\delta_\alpha G(\overline{\alpha})$ has the form

$$
\partial_\alpha G(\overline{\alpha}) = A(\overline{\alpha}) + Q_m(\overline{\alpha}) \ ,
$$

and the matrix $Q_m(\overline{\alpha})$ consists of the following elements:

$$
\begin{aligned}
Q_m^{kj}(\overline{\alpha}) &= \sum_{t\in\theta} \sum_{i=1}^{n} \frac{1}{m} \sum_{j=1}^{m} [x_t^{ij} - x_t^i(\overline{\alpha})] \Bigg(\langle b_t^i(\overline{\alpha}), \overline{\Gamma}\rangle^{-1} \frac{\partial}{\partial\alpha^j} \, a_t^{ik}(\overline{\alpha}) \\
&\quad - \frac{a_t^{ik}(\overline{\alpha})}{\langle b_t^i(\overline{\alpha}), \Gamma\rangle^2} \frac{\partial}{\partial\alpha^j} \langle b_t^i(\overline{\alpha}), \overline{\Gamma}\rangle \Bigg) \ .
\end{aligned}
\tag{8.20}
$$

As a result of this condition, equation (8.17) acquires the form

$$
\|A^{-1}(\overline{\alpha})Q_m(\overline{\alpha})\| < 1 \ .
$$

From expression (8.20), we conclude that if $m \to \infty$ and $\overline{\alpha} \to \alpha^*$, then elements of the matrix $Q_m(\overline{\alpha})$ tend to zero. Since

$$
E(x_t - x_t(\alpha^*)) = 0, \quad \forall t \in [0, T] \ ,
$$

therefore, with probability one,

$$
\lim_{m\to\infty} \frac{1}{m} \sum_{j=1}^{m} [x_t^{ij} - x_t^i(\alpha^*)] = 0 \ .
$$

This means that, at least for large values of m, there exists an α in the neighborhood of a point $\overline{\alpha}$ for which the iteration process in equation (8.14) converges. So we have proven the following.

Theorem 8.4

Let matrix $A(\overline{\alpha})$, which has elements as given in equation (8.18), be non-singular. Then there exists a sufficiently large value of m_1 such that for all $m > m_1$ there exists a sphere with radius δ_m in which the iterative process is correctly defined and converges to point $\overline{\alpha}$.

Proof

In addition to the above, see also Kalyaev and Zuev (1991).

It should be stressed that the limit of the iterative process $\overline{\alpha}$ depends on the sample size m. Take this into consideration by equating $\overline{\alpha}_m = \overline{\alpha}$. Let us show that $\overline{\alpha}_m \to \alpha^*$ as $m \to \infty$. With this objective, note that estimates $\overline{\alpha}_m$ and $\overline{\Gamma}_m$ satisfy the system of equations:

$$\sum_{t\in\theta}\sum_{i=1}^{n}\frac{1}{m}\sum_{j=1}^{m}\frac{x_t^{ij}-x_t^i(\overline{\alpha}_m)}{\langle b_t^i(\overline{\alpha}_m),\overline{\Gamma}_m\rangle}\ a_t^{ik}(\overline{\alpha}_m)=0$$

$$\sum_{t\in\theta}\sum_{i=1}^{n}\frac{1}{m}\sum_{j=1}^{m}\frac{b_t^{ik}(\overline{\alpha}_m)}{\langle b_t^i(\overline{\alpha}_m),\overline{\Gamma}_m\rangle}\left(1-\frac{[x_t^{ij}-x_t^i(\overline{\alpha}_m)]^2}{\langle b_t^i(\overline{\alpha}_m),\overline{\Gamma}_m\rangle}\right)=0\ , \qquad (8.21)$$

where $k = 1, 2, \ldots, L$.

These equations are transformed into the following:

$$\sum_{t\in\theta}\sum_{i=1}^{n}\frac{1}{m}\sum_{j=1}^{m}\frac{x_t^i(\alpha^*)-x_t^i(\overline{\alpha}_m)}{\langle b_t^i(\overline{\alpha}_m),\overline{\Gamma}_m\rangle}\ a_t^{ik}(\overline{\alpha}_m)$$

$$+\sum_{t\in\theta}\sum_{i=1}^{n}\frac{1}{m}\sum_{j=1}^{m}\frac{x_t^{ij}-x_t^i(\alpha^*)}{\langle b_t^i(\overline{\alpha}_m),\overline{\Gamma}_m\rangle}\ a_t^{ik}(\overline{\alpha}_m)$$

$$=F^k(\overline{\alpha}_m,\overline{\Gamma}_m,1/m)=0 \qquad (8.22)$$

$$\sum_{t\in\theta}\sum_{i=1}^{n}\frac{1}{m}\sum_{j=1}^{m}\frac{b_t^{ik}(\overline{\alpha}_m)}{\langle b_t^i(\overline{\alpha}_m),\overline{\Gamma}_m\rangle^2}\left(\langle b_t^i(\overline{\alpha}_m),\overline{\Gamma}_m\rangle\right.$$

$$\left.-\langle b_t^i(\alpha^*),\Gamma^*\rangle-[x_t^i(\alpha^*)-x_t^i(\alpha)]^2\right)$$

$$+\sum_{t\in\theta}\sum_{i=1}^{n}\frac{1}{m}\sum_{j=1}^{m}\frac{b_t^{ik}(\overline{\alpha}_m)}{\langle b_t^i(\overline{\alpha}_m),\overline{\Gamma}_m\rangle^2}\left(\langle b_t^i(\alpha^*),\Gamma^*\rangle\right.$$

$$\left.-[x_t^{ij}-x_t^i(\alpha^*)]^2+2[x_t^{ij}-x_t^i(\alpha^*)][x_t^i(\alpha^*)-x_t^i(\overline{\alpha}_m)]\right)$$

$$= F^{l+k}(\overline{\alpha}_m, \overline{\Gamma}_m, 1/m) = 0 \ ,$$

where $k = 1, 2, \ldots, L$.

Let us now take into account that, with probability 1, the following limit relations are fulfilled:

$$\lim_{m \to \infty} \frac{1}{m} \sum_{j=1}^{m} x_t^{ij} = x_t^i(\alpha^*)$$

$$\lim_{m \to \infty} \frac{1}{m} \sum_{j=1}^{m} [x_t^{ij} - x_t^i(\alpha^*)]^2 = \langle b_t^i(\alpha^*), \Gamma^* \rangle \qquad (8.23)$$

for all $t \in \theta$ and $i = 1, 2, \ldots, n$. Then, in the limit, the expressions in equation (8.22) have the following form:

$$F^k(\alpha, \Gamma, 0) = \sum_{t \in \theta} \sum_{i=1}^{n} \frac{x_t^i(\alpha^*) - x_t^i(\alpha)}{\langle b_t^i(\alpha), \Gamma \rangle} \ a_t^{ik}(\alpha) = 0$$

$$F^{l+k}(\alpha, \Gamma, 0) = \sum_{t \in \theta} \sum_{i=1}^{n} \frac{b_t^{ik}(\alpha)}{\langle b_t^i(\alpha), \Gamma \rangle^2} \Big(\langle b_t^i(\alpha), \Gamma \rangle$$
$$- \langle b_t^i(\alpha^*), \Gamma^* \rangle - [x_t^i(\alpha) - x_t^i(\alpha^*)]^2 \Big)$$

where $k = 1, 2, \ldots, L$. Obviously, vectors α^* and Γ^* are the solutions of the system: that is, $F(\alpha^*, \Gamma^*, 0) = 0$.

Theorem 8.5

Let the matrices $A(\alpha, \Gamma)$ and $B(\alpha, \Gamma)$ with elements

$$A^{kr} = \sum_{t \in \theta} \sum_{i=1}^{n} \frac{a_t^{ir}(\alpha^*) a_t^{ik}(\alpha^*)}{\langle b_t^i(\alpha^*), \Gamma^* \rangle}$$

$$B^{kr} = \sum_{t \in \theta} \sum_{i=1}^{n} \frac{b_t^{ik}(\alpha^*) b_t^{ir}(\alpha^*)}{\langle b_t^i(\alpha^*), \Gamma^* \rangle^2}$$

be non-singular. Then, with probability 1, the point (α^*, Γ^*) is a limit for $(\overline{\alpha}_m, \overline{\Gamma}_m)$, and there exists a sufficiently large value of m_1 such that, for $m > m_1$, equation (8.21) has a unique solution.

Proof

Denote the $2L \times 1$ vector consisting of the vector α and Γ by z: $z = (\alpha, \Gamma)$ and $\lambda = 1/m$. Note that $\overline{z}_m = (\overline{\alpha}_m, \overline{\Gamma}_m)$ satisfies both equations

$$F(\overline{z}_m, \lambda) = 0$$
$$F(\overline{z}_m, -\lambda) = 0$$

with $|\lambda| \leq 1$. In addition, let us define $F(z, \lambda)$ for the points where λ is not necessarily expressed by the ratio $1/m$, but where $\lambda \in (1/(m+1), 1/m)$:

$$F(z, \lambda) = F(z, 1/(m+1)) + [F(z, 1/m) - F(z, 1/(m+1))]$$
$$\cdot \{\lambda - [1/(m+1)]\}/\{(1/m) - [1/(m+1)]\} .$$

Using equation (8.23), one can show that, with probability 1,

$$\lim_{\lambda \to 0} [\det \partial_z F(z^*, \lambda)] = \det \partial_z F(z^*, 0)$$
$$= \det A(\alpha^*, \Gamma^*) \det B(\alpha^*, \Gamma^*) .$$

Consequently, the vector function $F(z, \lambda)$ continuously depends on its arguments (the matrix of its derivatives by z at the point $z = z^*$), $\lambda = 0$ is non-singular, and the following equality takes place: $F(z^*, 0) = 0$. Therefore, there exist some open neighborhoods S_1 and S_2, where $S_1 \subset R^{2L}$ and $S_2 \subset R^1$, of points z^* and $\lambda = 0$ such that for any $\lambda \in S_2$ equation $F(\bar{z}_m, \lambda) = 0$ has a unique solution, and $\bar{z}_m \to z^*$ for $\lambda = (1/m) \to 0$ with probability 1. In other words, for sufficiently large values of m, where $m \geq m_1$, one can always point out the neighborhood S_1 for (α, Γ) in which the function $\Phi_m(\alpha, \Gamma)$ is strictly convex, and the estimation problem has a unique solution which converges to true values α^* and Γ^* at $m \to \infty$.

It should be noted that this conclusion is based on a well-known theorem that is given below.

Theorem 8.6
Let the map $F : D \subset R^n \times R^p \to R^n$ be continuous in an open neighborhood $D_0 \subset D$ of point (x_0, y_0) for which $F(x_0, y_0) = 0$. Let the partial derivative $\partial_x F$, where $x \in R^n$, exist and be continuous in the neighborhood of (x_0, y_0) and let the matrix $\delta_x F(x_0, y_0)$ be non-singular. Then there exists some open neighborhoods, $S_1 \subset R^n$ and $S_2 \subset R^p$ of points x_0 and y_0, respectively, that for any $y \subset S_2$ the equation $F(x, y) = 0$ has a unique solution given by $x = Hy \in S_1$, and the map $H_1 S_2 \to S_1$ is continuous.

These results conclude our theoretical consideration of the parameter estimation problem. As mentioned above, its solution is intended to solve practical problems related to the data processing of observed data. In the next chapter we demonstrate the application of this methodology of mathematical modeling for the solution of a particular virus problem.

Chapter 9

An Experimental Influenza Investigation

The increasing study of realistic mathematical models in medicine is a reflection of their use in helping to understand a disease and in providing a practical health service. One of the main problems in this area is drug action. It is very important to have clear ideas about the strategy and tactics of drug use.

Here, we consider the mathematical model for the analysis of the action of drugs on influenza. There follows a short discussion about the mechanisms of anti-viral effects of chemical preparations and their prophylactic or therapeutic effects.

The first step in this problem is the construction of a mathematical model for a subsequent investigation of the possibility of a directed action in the viral process (using the influenza virus as an example). The model consists of a system of ordinary differential equations (ODEs). Its state variables describe the dynamics of the virus and of the cell populations of an animal organism, while the coefficients are interpreted as parameters of the corresponding interactions. Therefore, by determining the coefficients of the model from the results of experiments using chemical preparations possessing anti-viral action, it appears possible to judge how a drug affects the parameters of the process under study.

Of course, the construction of a mathematical model represents an idealization of the real scheme of the interaction of a virus with the host organism. Therefore, according to modern ideas (Ada and Jones, 1986; Polyak, 1987), the following main characteristics of the system must be considered: the number of viruses in the infected organ; the influence

155

of the interferon system on the process; and the cellular and humoral aspects of immunity.

In the biological experiments used in this chapter, in the main, standard methods of virological and immunological assays, which have been described in detail in Kalyaev (1987) and Marchuk *et al.* (1988), are used. The experimental model consists of F1 (CBA × C57Black) mice and the influenza virus A/PR8/34. Sublethal infection was brought about by the intranasal administration of 3.0–3.5 log EID_{50} of the virus and the investigations were carried out daily using not less than 3–5 animals each time. These experiments were conducted at the Institute of Experimental Medicine, St. Petersburg, in the virology laboratory (headed by Professor Polyak).

9.1   Model

The model presented here was constructed by D. Kalyaev (1987). All interactions between cellular and humoral components are consider as homogeneous reactions. Such factors as interferon (α, β, γ), immunoglobulin (M,G,A), and specific lymphocyte effectors are represented in the generalized sense. The variable "interferon" is used as a factor under the action of which a cell acquires an insensitivity to viral infection. "Antibodies" supply the specific neutralization of free viruses. "Specific lymphocyte effectors" are understood to be the population of lymphocyte effectors (without division into T and B subpopulations) with which the processes of antibody production and elimination of cells infected by viruses are connected. Let $C_v(t)$ be the number of cells infected by viruses; t be the time after infection (in days); $V(t)$ be the number of viral particles; $F(t)$ be the number of immunoglobulins; $L_e(t)$ be the number of specific effector lymphocytes, i.e., the end-product of differentiation from the lymphocyte precursors; and $L_p(t)$ be the number of lymphocytes having the corresponding specificity toward the viral antigen proliferating and differentiating under antigenic stimulus.

Suppose that the number of sensitive cells in the lungs is approximately constant (an acute but sublethal infection with $\alpha \approx 10\%$ infection of the pulmonary cells is being investigated) and $C_I \approx \alpha C_v$, where C_I is the number of cells that have become insensitive to infection under the action of interferon.

The conservation equations for C_v and V have the form

$$\frac{d}{dt}C_v = \underbrace{k_{C_v}V(C_0 - C_I - C_v)}_{I} - \underbrace{\gamma_{C_vL_e}C_vL_e}_{II} - \underbrace{\mu_{C_v}C_v}_{III}$$

$$\frac{d}{dt}V = \underbrace{k_vC_v}_{IV} - \underbrace{\gamma_{vF}VF}_{V} - \underbrace{\mu_v}_{VI} \tag{9.1}$$

where I corresponds to the appearance of new infected cells as the result of the interaction of the infected particles, V, and intact, susceptible cells $C_0 - C_I - C_V$: C_0 is a value characterizing the total number of cells potentially sensitive to the virus (mainly pulmonary epithelium). The term denoted by II gives the decrease in C_v due to specific lymphocyte effectors (the integral effect). Term III corresponds to the destruction of C_v cells due to the effect of the virus breaking down the cell in which it is being synthesized – a "background" cytotoxic effect not connected with specific lymphocyte effectors, etc. Term IV describes the synthesis of new viral particles and term V represents the neutralization of extracellular infective particles by immunoglobulins. Term VI represents the process of elimination of viral particles from the intracellular space, which is not caused by specific humoral factors; this includes the absorption and penetration of the virus into the cells and the loss of infectivity of the virus under the action of various types of "background" factors (temperature, inhibitors, acidity, etc.). It should be noted that the infective virus is in fact an effective stimulator of the immune system.

In the processes described above one can clearly distinguish both space and time hierarchies. The viral particle size of an animal cell is about 10^{-5}m (Libbert, 1982) and that of immunoglobulin is about 30–40×10^{-10}m. One viral particle is capable of infecting a cell (Luria et al., 1978), after which it can produce about 10^3 viral particles during the generation cycle of about 8 hours (after this cell dies) (Zhdanov, 1985). The time for absorption and penetration of the virus into the cell is about 10–20 minutes. The free virus disappears very fast.

The molecular reactions studied here (the binding of the virus by immunoglobulin and the absorption and penetration of viruses into new cells) have a characteristic time of the order of several minutes, while the complex cellular reactions leading to the destruction of the infected cells are characterized by larger times. So, neglecting the fast events with small characteristic times of the order of ten minutes to one hour, and using the chemical equilibrium law, we can write the approximation

$$V = \frac{k_v C_v}{\mu_v + \gamma_{vF} F} \ . \tag{9.2}$$

Then equation (9.1) acquires the form

$$\frac{d}{dt} C_v = \frac{k_{C_v} k_v C_v [C_0 - (1+v) C_v]}{\mu_v + \gamma_{vF} F} - \gamma_{C_v L_e} C_v L_e - \mu_{C_v} C_v \ . \tag{9.3}$$

At this point, we determine $C_v(0) = C_v^0$ as an initial condition. Here, C_v^0 is the effective number of infected cells from which infection begins [after 1–2 hours the transition processes cease and the dynamical equilibrium in equation (9.3) becomes valid]. For the description of the dynamics of immune system cells we adopt the following model scheme. The resting precursor cells (cellular phase G_0) specific to a given virus are activated by antigenic stimulus and transformed to proliferating cells, L_p, which in turn differentiate into terminal effector cells, L_e.

We distinguish these variables because there exists an hypothesis which says that the specific lymphocyte system plays a key role in the recovery of the organism. A number of mechanisms for virus neutralization and infected cell destruction are either directly determined by lymphocyte effectors (specific cytotoxic lymphocytes, immunoglobulin) or are controlled and induced by them (NK-cells, macrophages, K-cells) (Paul, 1984). In the model, the variables L_p and L_e are the main ones that reflect, on the whole, the defensive mechanism. We write the following conservation relation for L_p and L_e:

$$\frac{d}{dt} L_p = L_{p_0} \delta(t - t^*) + U(C_v)(\alpha_{L_p} L_p - \beta_{L_p} L_p)$$

$$\frac{d}{dt} L_e = U(C_v) \beta_{L_p} L_p - \mu_{L_e} L_e$$

$$U(C_v) = [1 - \exp(C_v/q)] \ . \tag{9.4}$$

In these equations, $L_{p_0} \delta(t - t^*)$ corresponds to the activation of resting precursors, L_p is their number, and t^* is the activation time. Here, $\delta(t)$ is the unit δ-function. It is supposed that proliferation and differentiation processes have a threshold character. They have a maximal rate at $C_v \gg q$, but cease when $C_v \ll q$.

The expression $-\mu_{L_e} L_e$ in equation (9.4) reflects the processes of activity loss and/or natural death of effector cells. The corresponding term in the first expression in equation (9.4) is absent since the variable

L_p must correspond to the formation of a clone of memory cells, the lifetime of which is 1–2 years (Asconas *et al.*, 1982). This is a very long time interval in comparison with that needed for the acute process development. For antibodies F we have the usual equation,

$$\frac{d}{dt}F = \rho L_e - \gamma_{FV} FV - \mu_F F \;,$$

and, using equation (9.2), we obtain

$$\frac{d}{dt}F = \rho L_e - \frac{\gamma_{Fv}k_v FC_v}{\mu_v + \gamma_{vF} F} - \mu_F F \;. \tag{9.5}$$

From equation (9.2), it can be seen that

$$\ln V = \ln\left(\frac{k_v}{\mu_v}C_v\right) - \ln\left(1 + \frac{\gamma_{vF}}{\mu_v}F\right) \;,$$

where the second term is small in comparison with the first one. So in further analysis we adopt

$$V = \frac{k_v}{\mu_v}C_v \;, \tag{9.6}$$

but dependence on F is represented in the right-hand sides of equations (9.3) and (9.5).

Let us use equation (9.6) to write the model equations in the form that was used for influenza dynamics:

$$\begin{aligned}
\frac{d}{dt}V &= \frac{\alpha_1 + \alpha_{10}}{1 + \alpha_9 F}V - \alpha_2 V^2 - \alpha_3 V L_e - \alpha_{10}V \\
\frac{d}{dt}L_p &= L_{p0}\delta(t - t^*) + \alpha_{11}L_p[1 - \exp-(\alpha_{12}V)] \\
\frac{d}{dt}L_e &= \alpha_4 L_p[1 - \exp(\alpha_{12}V)] - \alpha_5 L_e \\
\frac{d}{dt}F &= \alpha_6 L_e - \frac{\alpha_7 VF}{1 + \alpha_9 F} - \alpha_8 F \\
V(0) &= V^0, \quad L_p(0) = L_e(0) = 0, \quad F(0) = 0 \;.
\end{aligned} \tag{9.7}$$

In such a form the system can be used in practical work because in the experimental system we estimate $V(t)$ rather than $C_v(t)$.

In the case studied here (acute, uncomplicated viral infection), one can assume that there is a definite scenario in which defensive mechanism action is realized. So, as shown by Kalyaev (1987), the structure of the solutions of the model corresponding to the process being studied is such that the interferon system $(\alpha_2 V^2)$ and the generalized cytotoxic action of effectors are the mechanisms that supply saturation in virus dynamics (during 4–5 days), which is followed by a sharp decrease in the number of virus infections in the lungs (5–8 days).

The influence of the variable F on the solution is practically absent. It begins to manifest itself later (8–14 days) when F reaches large values. Until this moment, the number of viruses is quite small in comparison with the maximum. So the term $\alpha_2 V^2$ is negligibly small for this interval. It is likely that such a solution structure reflects the real picture of the defense mechanism over time. Due to the speculations mentioned above, equation (9.7) differs in structure from (9.3). In the term $\alpha_2 V^2$ the dependence on F is absent. It presents itself only in the

$$\frac{\alpha_1 + \alpha_{10}}{1 + \alpha_9 F} V$$

linear term. The model derived matches the experimental data quite well.

9.2   Data Processing

It is clear that the values of the coefficients of the model determine the dynamics of the "infection" in the model. In the final account, the outcome of the disease depends on them. As was considered in Chapters 6 and 7, the possibility of direct action on the course of the disease with the aid of drugs is of significant interest. Within the framework of our model such actions are interpreted as changes in the vector coefficient α.

Therefore, in order to understand how a drug affects the process under study, the inverse problem must be solved, i.e., from the dynamics of the variables of the model observed in the experiment it is necessary to calculate the coefficients. As a result of the solution of this problem from the results of experiments with animals that have and have not received the drug, it is possible to understand the mechanism of the drug's action on the process. The method for the problem's solution (with the aid of which the results given below were obtained) is described in Chapters 7 and 8, and takes into account features of the model and the results of immunological experiments.

In *Figure 9.1*, the solutions of the model in equation (9.7) are shown by solid lines. The solutions correspond to coefficients calculated from data for an animal group that had not been given the drug. The observed

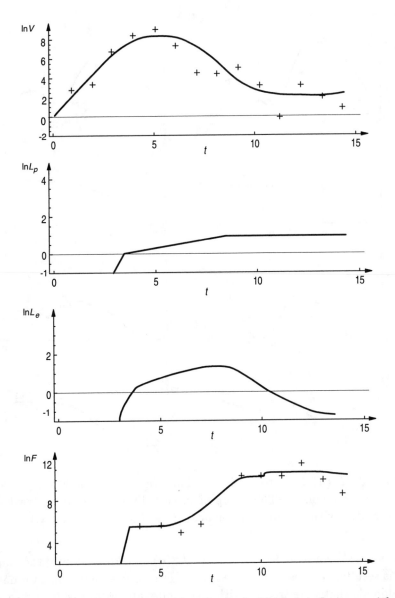

Figure 9.1. Control group of animals (sublethal influenza without treatment). The solution of the model is given by solid lines. Plus signs (+) denote experimental data. The term ln V is a virus in a dose of 50% embryonal infection for animal lungs. The term ln F is the amount of (IgM + IgG) in the serum.

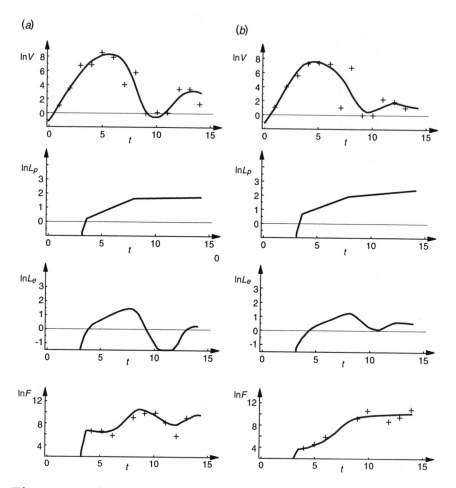

Figure 9.2. (*a*) The group of sublethally infected animals which were treated with ionol. (*b*) The group of sublethally infected animals which were treated with ε-aminocaprioc acid. (See *Figure 9.1* for key.)

results are shown by pluses. In *Figure 9.2(a)* and *Figure 9.2(b)* the analogous results are presented. These were derived from experimental data on animals that were treated with ionol and ε-aminocaproic acid respectively. The calculated values of coefficients are given in *Table 9.1*. The application of the verification criterion of a statistical hypothesis (Section 7.2) shows that changes in the coefficient vector under treatment action are significant and have a high level of statistical validity. The

Table 9.1. Maximum-likelihood estimates of coefficients.

Coefficients	Figure 9.1	Figure 9.2(a)	Figure 9.2(b)
α_1	2.0	2.328	2.581
α_2	0.0	0.0	4.316×10^{-4}
α_3	0.958	0.972	0.950
$\alpha_4{}^*$	1.8	1.8	1.8
α_5	0.891	1.51	1.0
α_6	1.530×10^4	2.178×10^4	1.411×10^3
$\alpha_7{}^*$	2.190×10^{-2}	2.190×10^{-2}	2.190×10^{-2}
α_8	0.0	1.834	0.0
α_9	4.730×10^{-6}	2.226×10^{-5}	0.0
$\alpha_{10}{}^*$	7.0	7.0	7.0
α_{11}	0.201	0.331	0.294
α_{12}	3.870×10^{-3}	1.115×10^{-2}	7.664×10^{-2}
$t*$	3 days	3 days	3 days
V^0	5.037×10^{-2}	-1.139	-1.394
L_{p0}	1	1	1

conclusions below are to some extent hypotheses and serve as illustrations of the methodology for the solution of such problems.

It is interesting that for the third group (with ε-aminocaproic acid) α_6, which characterizes the rate of antibody production, is one order smaller in comparison with the control group. For the same group we derived $\alpha_3 = 0$. We can conclude from this that the influence of immunoglobulin on the process under study is weak in comparison with other defense mechanisms. In the case of the third group, $\alpha_2 \neq 0$, while in the cases of the second and the first groups, $\alpha_2 = 0$. This can be interpreted as the fact that in the case of the first and second groups the maximum of the virus curve and its decrease is mainly explained by the cytotoxic mechanism mediated by lymphocyte effectors. In the third group, added to this is the saturating mechanism which is due to the action of interferon. Except for this, the initial value V_0 decreases for those groups which were treated. In the model in equation (9.7) the term V_0 is the number of viruses which reach the lung cells after intranasal infection ($V_0 < 3.0 - 3.5 \log \text{EID}_{50}$). This fact allows us to formulate a hypothesis about the prophylactic effect of these drugs.

As we can see from *Table 9.1*, in three groups of animals the coefficients α_1, α_3, and α_5 are practically unchanged. Also, such process parameters as the general ototoxic effect of effectors, the rate of virus multiplication, and the rate of removal (and inactivation) of effector cells have not changed.

For animal groups treated with drugs, α_{12} is several times smaller than in the control group. This might reflect the increase in the virus population threshold at which the proliferation and differentiation processes possess the maximal rate. The studies show that these changes in coefficients are very essential. The fact that the coefficient α_{11} increases 1.5 times indicates that there are more intensive processes of proliferation and of immune memory formation. Perhaps this is a reason for the observed anti-viral effects of drugs in this study.

The results obtained confirm the possibility of the application of an advanced approach to the study of the action of drugs on the acute infectious process. Beyond this, the mathematical model enables us to obtain quantitative characteristics of the internal process using in vivo data. A larger experimental database enables us to create and to test a new scenario of infectious disease development.

Chapter 10

Parameter Estimation for a Given Organism

In this chapter the problem of estimation of model parameters for a given organism when the number of observed data is less than the number of unknown parameters is studied.

10.1 Motivation

The previous discussion in Chapters 7 and 8 demonstrates the efficiency of data analysis on the basis of mathematical models. It has been assumed that every trajectory from the set

$$X_m = \{x_t^i, t \in \theta^i, i = 1, 2, \ldots, m\}$$

is a realization defined on θ^i of a random process

$$\tilde{x}_t = \{\tilde{x}_t(\omega), \ t \in [0, T], \ \omega \in \Omega\}$$

such that

$$E\tilde{x}_t = \int_\Omega \tilde{x}_t(\omega) dP(\omega) = x_t(\alpha^*) \ ,$$

where $x_t(\alpha^*)$ is a solution of the model

$$\frac{d}{dt} x_t = f(x_t, \alpha^*), \ t \in [0, T], \ x_0 = c \ . \tag{10.1}$$

By the same token, we neglect the individual character of trajectories and the estimation $\overline{\alpha}_m$ of α^* is an average value for a given group of patients (organisms). This is sufficient for the solution of the problem of parameter estimation which was mentioned in Chapter 9. At the same time there are the practical problems: we need to know the individual characteristics of the process in question, for example, the investigation of patient disease dynamics and individual drug management. For this purpose, the approach which is mentioned in the previous chapters is not sufficient.

For the individual estimation of model parameters, we must have at least L measurements of the state vector in $t_1, t_2, \ldots, t_L$, where L is the number of unknown parameters. If we had the possibility of taking measurements every day (a dream) we could evaluate the parameters after L days. Since $L \sim 10$, our individual forecast of disease course loses its significance, because it is very important to make a forecast at the beginning of the disease. Here we discuss one approach to solving this problem which is based on the idea that the parameters of the process are connected by means of a common factor. The solution of this problem may be based on the idea of the model parametrization which was proposed by Marchuk in the 1980s and significantly developed by Pogozhev (1988, 1989). Formalization of this idea from the identification point of view is given by Asachenkov et al. (1992).

Consider two organisms, the first one we will call "basal" and a second one "studied" or "given". Suppose that the parameters of the basal organism are known. Consequently we have two models

$$
\begin{aligned}
\frac{d}{dt}\underline{x}_t &= f(\underline{x}_t, \underline{\alpha}), \ \ t \in [0, T] \\
\underline{x}_0 &= \underline{c}, \ \underline{x}_t \in \mathbf{R}^n, \ \underline{\alpha} \in \mathbf{R}^L
\end{aligned}
\tag{10.2}
$$

and

$$
\begin{aligned}
\frac{d}{dt}x_t &= f(x_t, \alpha), \ \ t \in [0, T] \\
x_0 &= c, \ x_t \in \mathbf{R}^n, \ \alpha \in \mathbf{R}^L \ .
\end{aligned}
\tag{10.3}
$$

The first one describes a disease process in the basal organism, where $\underline{\alpha}$ is a known vector of parameters, and the second describes a disease process in the studied one, where α is unknown.

Assumption 10.1.

Let parameters α and $\underline{\alpha}$ be connected by the correlation of similarity

$$
\alpha = v(\underline{\alpha}, HL) \in \mathbf{R}^L, \ \ \underline{\alpha} \in \mathbf{R}^L, \ \ HL \in \mathbf{R}^1
$$

where HL is a coefficient of similarity and $v(x, y)$ is a vector function such that $v(\underline{\alpha}, 1) = \underline{\alpha}$.

Now, if the function $v(x, y)$ and a vector $\underline{\alpha}$ are known, then for estimating the unknown parameter α we must evaluate only the coefficient of similarity. Clearly, the estimation of one unknown parameter is a simpler problem.

Explanation

The state variables of the disease model are the concentrations of interacting particles (cells, molecules, etc.). These particles interact with each other through the movement of the liquid media of the organism (blood, lymph). In this way, the micromovement intensity of particles in the liquid media is an important factor influencing the rate of biochemical processes in the body. Suppose that the model parameters are connected by means of some common factors denoted by G such that

$$\alpha = \tilde{v}(\underline{\alpha}, G) \in \mathbb{R}^L \ ,$$

where $\underline{\alpha}$ is a known vector for the so-called basal organism or "average" for the given group of patients (for the evaluation of this parameter we can use the methods mentioned above) and G is the unknown micromovement intensity of interacting particles.

Let $\underline{G}$ denote the micromovement intensity of interacting particles for the basal organism, then

$$\alpha = \tilde{v}(\underline{\alpha}, G) = v(\underline{\alpha}, HL), \quad HL = \frac{G}{\underline{G}} \ .$$

So, we can interpret the parameter HL as the normalized intensity of the micromovement of interacting particles in the liquid media of the organism. This means that for M patients we have to write M models,

$$\frac{d}{dt} x_t^i = f(x_t^i, v(\underline{\alpha}, HL^i)), \quad x_0^i = c^i$$
$$i = 1, 2, \ldots, M \ ,$$

which are distinguished only by the parameter HL.

10.2   Individual Parameter Estimation

Consider the model

$$\frac{d}{dt}x_t = f(x_t, \underline{\alpha}, HL), \quad x_0 = c, \quad t \in [0, T] \ ,$$

where $\underline{\alpha}$ is a known vector, and the health level HL is an unknown individual parameter.

To solve this problem we use the approach suggested in the previous chapters. The stochastic model for the description of the real trajectories has the form

$$\frac{d}{dt}x_t^\varepsilon = f(x_t^\varepsilon, \overline{\alpha} + \xi_{t/\varepsilon}, HL), \quad x_t^\varepsilon = c, \quad t \in [0, T] \ .$$

Suppose that the true value is HL^*, given by $HL^* = HL_k + \delta HL_k$, where δHL_k is small. The analysis of real data shows that the deviation of HL in the population is approximately 20%. According to Chapter 8, for $\delta x_t = x_t^\varepsilon - x_t(HL_k)$ we can write the model

$$\begin{aligned}
\frac{d}{dt}\delta x_t &= f_x(x_t^k, \underline{\alpha}, HL_k)\,\delta x_t + f_{HL}(x_t^k, \underline{\alpha}, HL_k)\,\delta HL \\
&\quad + f_\alpha(x_t^k, \underline{\alpha}, HL_k)\frac{d}{dt}w_t
\end{aligned}$$

$$\delta x_0 = 0, \quad \mathrm{cov}\,(w_t, w_t) = \Gamma t \ .$$

Here, x_t^k is a solution of

$$\frac{d}{dt}x_t^k = f(x_t^k, \underline{\alpha}, HL_k), \quad x_0^k = c, \quad t \in [0, T] \ .$$

Let us suppose that the observations are

$$x_t = Qx_t^\varepsilon + \eta_t, \quad t \in \Theta = \{t_1, t_2, \ldots, t_N\}$$

where η_t from R^r is observation noise independent of x_t^ε, $E\eta_t = 0$, $\mathrm{cov}\,(\eta_t, \eta_t) = \Gamma_0$, $t \in \Theta$, and Q is an $r \times (r \le n)$ matrix. For example, Q may be diagonal with unit elements, which means that we measure only r components of the state vector. For $s < t$ we can write

$$\begin{aligned}
\delta x_t &= R(t, s)\,\delta x_s + \delta HL R(t, s)\int_s^t R^{-1}(\tau, s)f_{HL}(x_\tau^k, \underline{\alpha}, HL_k)\,d\tau \\
&\quad + R(t, s)\int_s^t R^{-1}(\tau, s)f_\alpha(x_\tau^k, \underline{\alpha}, HL_k)\,dw_\tau
\end{aligned}$$

where $R(\,.\,)$ is a solution of the equation

$$\frac{d}{dt}R(\tau,s) = f_x(x_\tau^k, \underline{\alpha}, HL_k)R(\tau,s)$$

$$\tau \in [s,t], \quad R(s,s) = I \ .$$

Let us denote

$$A(t,s) = R(t,s)\int_s^t R^{-1}(\tau,s)f_{HL}(x_\tau^k, \underline{\alpha}, HL_k)\,d\tau$$

$$W_t = R(t,s)\int_s^t R^{-1}(\tau,s)f_\alpha(x_\tau^k, \underline{\alpha}, HL_k)\,dw_\tau \ .$$

Using this notation for $t = t_i$ and $s = t_i - 1$, we have

$$\begin{aligned}
\delta x_i &= R_i\delta x_{i-1} + A_i\delta HL + W_i \\
x_i &= Q(\delta x_i + x_i^*) + \eta_i \\
i &= 1,2,\dots \ ,
\end{aligned} \tag{10.4}$$

where $R_i = R(t_i, t_{i-1})$, $A_i = A(t_i, t_{i-1})$, $W_i = W_{t_i}$, $\delta x_i = \delta x_{t_i}$, $x_i = x_{t_i}$, and $x_i^k = x_{t_i}^k$.

Now we have to solve the corresponding filter problem. For this purpose we construct the likelihood ratio

$$L(X, \delta HL | HL_k) = \frac{P(X, \delta HL | HL_k, H_1)}{P(X | H_0)} \ . \tag{10.5}$$

The estimation of δHL_k is defined by the condition

$$\max_{\delta HL} L(X, \delta HL | HL_k) = L(X, \delta HL_k | HL_k) \ .$$

Here, $P(X, \delta HL | HL_k, H_1)$ is a conditional joint density corresponding to the hypothesis H_1, according to which the observations, given by $x_i = Q(\delta x_i + x_i^k) + \eta_i$, contain signal and noise. The term $P(X | H_0)$ is a conditional density which corresponds to the hypothesis H_0, according to which the signal is absent in the observations $x_i = \eta_i$. For $P(X | H_0)$ we have

$$P(X | H_0) = \prod_{i=1}^N p(x_i)$$

where

$$p(x) = [(2\pi)^r \det(\Gamma_0)]^{-1/2}\exp\left(-x^T\Gamma_0 x\right) \ .$$

To write the density $P(X, \delta HL | HL_k, H_1)$ let us recall (Balakrishnan, 1984) that the random values

$$
\begin{aligned}
y_1 &= x_1 \\
y_2 &= x_2 - E[x_2|x_1] \\
y_3 &= x_3 - E[x_3|x_2, x_1] \\
&\vdots \\
y_N &= x_N - E[x_N|x_{N-1}, \ldots, x_1]
\end{aligned}
$$

are independent. Therefore,

$$
P(X, \delta HL | HL_k, H_1) = \prod_{i=1}^{N} p_i(y_i)
$$

where the densities are Gaussian.

Now we have to define the conditional expectations

$$
E[x_i|X_{i-1}], \quad X_{i-1} = \{x_{i-1}, x_{i-2}, \ldots, x_1\}
$$

$$
i = 2, 3, \ldots, N \ .
$$

From equation (10.4) we have

$$
y_i = x_i - E(x_i|X_{i-1}) = x_i - Q(\delta \hat{x}_i + x_i^k) \ ,
$$

where

$$
\delta \hat{x}_i = E(\delta x_i|X_{i-1})
$$

is an estimate of the signal δx_i on the set X_{i-1}. To find this estimate we can use the filter equation (Balakrishnan, 1984)

$$
\begin{aligned}
\delta \hat{x}_{i+1} &= R_i \delta \hat{x}_i + \delta HLA_i + R_i P_i Q^T \left[Q P_i Q^T + \Gamma_0 \right]^{-1} \\
&\quad \times \left[x_i - Q(\delta \hat{x}_i + x_i^k) \right]
\end{aligned}
$$

where

$$
\begin{aligned}
P_{i+1} &= (R_i - K_i Q) P_i (R_i - K_i Q)^T + \Gamma_i + K_i \Gamma_0 K_i^{-1} \\
K_i &= R_i P_i Q^T [Q P_i Q^T + \Gamma_0] \\
P_0 &= \Gamma_0 \\
\delta \hat{x}_0 &= 0 \ .
\end{aligned}
$$

Now we are ready to calculate the likelihood ratio of equation (10.5) to estimate δHL. Clearly, this expression depends on δHL explicitly. This means that when the matrices Γ and Γ_0 are known we can obtain the explicit formulae for δHL_k by differentiating on δHL and making the derivative tend toward zero. Therefore, we have the iterative process for the HL calculation:

$$HL_{k+1} = HL_k + \delta HL_k, \quad k = 0, 1, \ldots .$$

If both of the matrices Γ and Γ_0 are unknown, as well as HL, then estimating HL is a more complicated problem.

Finally, let us discuss the case when $\underline{\alpha}$ is also unknown. To this end we need the set X_m of trajectories obtained on the group of m patients. According to our approach every trajectory

$$
\begin{aligned}
X^i &= \{X_t^i, t \in \Theta, \Theta\} \\
\Theta &= \{t_1, t_2, \ldots, t_N\}
\end{aligned}
$$

is described by the model

$$\frac{d}{dt} x_t^i = F(x_t^i, \underline{\alpha}, HL^i)$$

$$x_t^i \in \mathbb{R}^n, \quad \underline{\alpha} \in \mathbb{R}^L, \quad HL^i \in \mathbb{R}^1, \quad i = 1, 2, \ldots, m . \tag{10.6}$$

Denote the $mn \times 1$ vector of state variables by,

$$y_t = \begin{pmatrix} x_t^1 \\ x_t^2 \\ \vdots \\ x_t^m \end{pmatrix}$$

and then

$$F(y_t, \underline{\alpha}, HL) = \begin{pmatrix} f(x_t^1, \underline{\alpha}, HL^1) \\ \vdots \\ f(x_t^m, \underline{\alpha}, HL^m) \end{pmatrix}$$

is a vector function of the right-hand side. Now we can write equation (10.4) in the form

$$\frac{d}{dt} y_t = F(y_t, \underline{\alpha}, HL) ,$$

where HL is a vector with the components HL^1, $HL^2, \ldots, HL^m$. Here we have $L + m$ unknown parameters. At the same time the number of measurements is Nm. To estimate the unknown parameters the following inequality must be valid:

$$Nm > L + m$$

or

$$N > \frac{L}{m} + 1 \ .$$

Certainly, if $i \neq j$, $HL^i \neq HL^j$. So, introducing a personal parameter into the model allows us to reduce the number of measurements.

To construct the iterative algorithm let us write the perturbed model

$$\frac{d}{dt} y_t^\varepsilon = \tilde{F}(y_t^\varepsilon, \underline{\alpha}_0 + \delta\alpha, \xi_{t/\varepsilon}, HL_0 + \delta HL) \ ,$$

where

$$\tilde{F}(y_t, \underline{\alpha}, \xi_t, HL) = \begin{pmatrix} f(x_t^1, \underline{\alpha} + \xi_t(\omega_1), HL^1) \\ \vdots \\ f(x_t^m, \underline{\alpha} + \xi_t(\omega_m), HL^m) \end{pmatrix} \ .$$

Here, the terms given by $\xi_t(\omega^i)$, where $i = 1, 2, \ldots, m$, are independent realizations of the random process

$$\{\xi_t(\omega), t \in [0, T], \omega \in \Omega\} \ .$$

Now we can use the algorithm which was mentioned above. For the estimation of HL we have an iteration process again. However, the analytical form for $v(x, y)$ is still an unknown. This suggests that we should study the process in question at the microlevel.

10.3   The System of Interacting Particles

Consider a system of interacting particles in the liquid media of the organism. Let, for the time t,

$$q(t) = \left(q^1(t), q^2(t), \ldots, q^s(t) \right)^T$$

denote the generalized coordinates (Bukhgolts, 1972) of the interacting particle system and let

$$\frac{d}{dt} q(t) = \{\dot{q}_i(t), \quad i = 1, 2, \ldots, s\}$$

denote the corresponding generalized rates, where s is the number of degrees of freedom in the system of interacting particles. Using the generalized coordinates allows us, in principle, to describe the complex, progressive, and revolving movement of the particles in the liquid media (blood and lymph). The number of degrees of freedom, s, is of the order of approximately 10^7 (the number of particles).

Now, we can write, formally, the movement equation in the form

$$\frac{d}{dt} q(t) = \psi(\pi(t), \xi_{t/\varepsilon}) , \quad t \geq 0 , \quad q(0) = q_0 ,$$

where ψ is a generalized force which describes the disturbance of the liquid media by the organs and tissues of a living body; $\pi(t)$ is a disturbance with a characteristic time given by $\tau_1 \approx 1$ second which describes the movement of the liquid media due to a person's heart beat. It should be mentioned that the variables $q(t)$ and $\pi(t)$ have the same characteristic times; ξ_t/ε is a disturbance with a characteristic time τ_0 given by $\tau_0 \approx 10^{-3}$ seconds (the average time interval between particle contacts); and $\varepsilon > 0$ is a small parameter of the order of $\tau_0/\tau_1 \sim 10^{-3}$.

The movement of the system due to $\pi(t)$ we will call a trend and movement due to ξ we will call a random movement. To describe a trend we need to average over the fast variable. Suppose that the following limit exists:

$$\overline{\psi}(\pi) = \lim_{T\delta\infty} \frac{1}{T} \int_0^T \psi(\pi, \xi_t) \, dt .$$

More precisely, the following limit holds, uniformly by τ, for any $\delta > 0$ and $T > 0$:

$$\lim_{T\delta\infty} P\left\{ \left| \frac{1}{T} \int_\tau^{\tau+T} \psi(\pi, \xi_t) dt - \overline{\psi}(|pi) \right| > \delta \right\} = 0 ,$$

where $P\{A\}$ is the probability of event A. Thus we have

$$\frac{d}{dt} \overline{q}(t) = \overline{\psi}(\pi(t)) .$$

Consider the normalized difference

$$\frac{1}{\sqrt{\varepsilon}}[q(t) - \overline{q}(t)] = \frac{1}{\sqrt{\varepsilon}} \int_0^T \left[\psi(\pi(s), \xi(s/\varepsilon)) - \overline{\psi}(\pi(s)) \right] ds .$$

Theorem 10.1

Assume that the random process ξ_t with values in the set $\mathbb{R}^s$ has a piecewise continuous trajectory, with probability 1, and satisfies the condition of strong mixing with the coefficient $\gamma(\tau)$ such that

$$\int_0^\infty \tau[\gamma(\tau)]^{1/5}\,d\tau < \infty$$

and

$$\sup_{x,t} E|\psi(x,\xi_t)|^3 < C_1 < \infty \ .$$

Moreover, for all $\tau \in [0,T]$ and $C < \infty$,

$$\left|\int_0^\tau [E\psi(\pi(s),\xi(s/\varepsilon)) - \overline{\psi}(\pi(s))]\,ds\right| < C\varepsilon \ .$$

Then the process

$$\zeta_t^\varepsilon = \frac{1}{\sqrt{\varepsilon}}\,(q(t) - \overline{q}(t))$$

as $\varepsilon \to 0$ weakly converges on $[0,T]$ to a Gaussian process with independent increments, zero mathematical expectation, and a covariance matrix given by

$$R = \{R^{ij}(t)\}$$
$$R^{ij} = \int_0^t A^{ij}(\pi(s))\,ds \ ,$$

where

$$A^{ij}(x) = \lim_{T\delta\infty} \frac{1}{T}\int_0^T \int_0^T A^{ij}(x,s,t)\,ds\,dt$$
$$A^{ij}(x,s,t) = E\left[\psi^i(x,\xi_s) - E\psi^i(x,\xi_s)\right]$$
$$\times \left[\psi^j(x,\xi_t) - E\psi^j(x,\xi_t)\right] \ .$$

Using this theorem, for small ε, we can rewrite the movement equation in the form

$$\frac{d}{dt}\,q(t) = \overline{\psi}(\pi(t)) + \sqrt{\varepsilon}B(t)\frac{d}{dt}\,w_t \tag{10.7}$$

where w_t is a Wiener process and $B(t)$ is a matrix such that

$$B(t)B^T(t) = R(t) \ .$$

It is well known (see, for example, Gardiner, 1984), that for the diffusion process x_t,

$$dx_t = A(x,t)dt + B(x,t)\,dw_t \ ,$$

the conditional probability density $p(x,t|x_0,t_0)$ satisfies the Fokker–Planck equation

$$\frac{\partial}{\partial t} p(x,t) = -\sum_i \frac{\partial}{\partial x_i} [A_i(x,t)p(x,t)]$$

$$+ \frac{1}{2} \sum_{i,j} \frac{\partial^2}{\partial x_i \partial x_j} [D_{ij}(x,t)p(x,t)]$$

$$p(x,t) = p(x,t|x_0,t_0)$$
$$p(x,0) = \delta(x - u(0)) \ ,$$

where A is a shift vector, and $D = BB^T$ is a diffusion matrix.

Thus, for the system in equation (10.7) we have

$$\frac{\partial}{\partial t}p(q,t) = -\sum_i \overline{\psi}(\pi(t))\frac{\partial p(q,t)}{\partial q_i} + \frac{\varepsilon}{2}\sum_{i,j} R_{ij}\frac{\partial^2 p(q,t)}{\partial q_i \partial q_j} \ ,$$

In this way, in the system of interacting particles the motion of the particles together with the liquid media due to the heart beating, etc., is described by a vector of shift and the Brownian movement by the covariance matrix $R(t)$. It should be noted that for the centralized process

$$u(t) = q(t) - \overline{q}(t)$$

and we have

$$\frac{\partial}{\partial t}p(u,t) = \frac{\varepsilon}{2}\sum_{i,j} R_{ij}\frac{\partial^2 p(u,t)}{\partial u_i \partial u_j} \ .$$

Now consider the characteristic time τ_2 during which the separate micromovements of particles in the liquid media of the organism must be thoroughly mixed. This is approximately 1 minute. It means that for a time interval greater than τ_2 we can consider our stochastic process as a stationary stochastic process and

$$\varepsilon R(t) = D \ ,$$

where the matrix D is not dependent on time. Under this condition we can start to study the micromovement similarity for two organisms.

10.4   Similarity of the Diffusion Processes

Definition 10.1
Two diffusion processes $x(t)$ and $y(t)$ are stochastically equivalent
if, for all t, their conditional densities $p_x(u,t)$ and $p_y(u,t)$ satisfy

$$p_x(u,t) = p_y(u,t)$$

for almost all u.

Compare two organisms. The first is a so-called basal one, the pa-
rameters of which are known: $\underline{\alpha} = (\underline{\alpha}^1, \underline{\alpha}^2, \ldots, \underline{\alpha}^L)^T$. The parameters
of the second organism, $\alpha = (\alpha^1, \alpha^2, \ldots, \alpha^L)^T$, are unknown. Now we
want to investigate its similarity. Let, for $t > \tau_2$,

$$q(t) = \{q_i(t), i = 1, 2, \ldots, s\}$$
$$\underline{q}(t) = \{\underline{q}_i(t), i = 1, 2, \ldots, s\}$$

be generalized coordinates, and

$$\frac{d}{dt} q(t) = \left\{ \frac{d}{dt} q_i(t), \ i = 1, 2, \ldots, s \right\}$$
$$\frac{d}{dt} \underline{q}(t) = \left\{ \frac{d}{dt} \underline{q}_i(t), \ i = 1, 2, \ldots, s \right\}$$

be the generalized rates of the systems of interacting particles, where
underlined vectors correspond to the basal organism.

Pogozhev's Hypothesis
From the thorough analysis undertaken by Shmidt-Nielson (1987),
it follows that the general, vital functions are approximately similar,
not only in different bodies but in many mammals as well. Organisms are much alike: they consist of cells of approximately equal
size. However, their total number defines each body size. The sizes
of intercellular space, blood capillaries, erythrocytes, lymphocytes,
macrophages, and other particles interacting in the immune process
are also almost the same. So is the volume fraction of an organ's
liquid media such as blood plasma, lymph, and intratissue fluid, its
temperature and viscosity, and the concentration of lymphocytes,
proteins, glucose, and other interacting particles. Vital lung volume
and heart mass vary with body mass: about five systoles fall on one
breathing cycle in humans and many other mammals.

The analysis of these and other modern physiological data allows us to adopt the following assumption of similarity of interacting particle micromovements in the liquid media of the organisms to be compared:

$$\frac{d}{dt}q(t) \doteq \frac{V_b}{\underline{V}_b}\frac{d}{dt}\underline{q}\left(\frac{t\underline{\tau}_c}{\tau_c}\right) , \tag{10.8}$$

where V_b and $\underline{V}_b$ are specific rates of blood circulation (calculated per mass unit) and τ_c and $\underline{\tau}_c$ are average durations of the cardiac cycle of the organism in question and the basal one, respectively. The symbol for "equivalence" ($\doteq$) in equation (10.8) and below is interpreted as the stochastic equivalence of a corresponding random process.

Stochastic equivalence means that all available statistical information about the organism in question, which we can obtain in terms of generalized rates $dq(t)/dt$, we can recalculate using generalized rates for the basal organism. For this purpose we must multiply a generalized rate of the basal organism by the ratio $V_b/\underline{V}_b$ and change the time scale $\underline{\tau}_0/\tau_0$. In other words, for a particular body we use its individual "physiological" time. This fact corresponds to the main theses put forward by Shmidt-Nielson (1987).

Let us now return to the idea of parametrization and consider the relationship

$$u(t) \doteq a\underline{u}(bt) ,$$

where a and b are unknown constants, such that $a, b \in R^1$. To find them, let us consider the process

$$\tilde{u}(t) = a\underline{u}(bt)$$

for times greater than τ_2 and establish the condition of stochastic equivalence $u(t) \doteq \tilde{u}(t)$. The conditional densities for $u(t)$ and $\underline{u}(t)$ are

$$\frac{\partial}{\partial t}p(u,t) = \frac{1}{2}\sum_{i,j}D_{ij}\frac{\partial^2 p(u,t)}{\partial u_i \partial u_j}$$

$$\frac{\partial}{\partial t}\underline{p}(u,t) = \frac{1}{2}\sum_{i,j}\underline{D}_{ij}\frac{\partial^2 \underline{p}(u,t)}{\partial u_i \partial u_j} . \tag{10.9}$$

Now, we can write the analogous equation for the process $\tilde{u}(t)$. Clearly,

$$\tilde{p}(u,t) \;=\; \underline{p}(au,bt)$$

$$\tilde{p}(u,t) = \underline{P}(ax \le u,t) \;=\; \underline{P}(x \le u/a,t) = \underline{P}(u/a,t) \;.$$

Then, from (10.9) we have

$$\frac{\partial}{\partial t}\,\tilde{p}(u,t) = \frac{1}{2}\sum_{i,j} a^2 b\underline{D}_{ij}\,\frac{\partial^2 \tilde{p}(u,t)}{\partial u_i \partial u_j} \;.$$

Using equation (10.9) and definition 10.1, for stochastic equivalence of the processes $u(t)$ and $\tilde{u}(t)$ the following relation must hold:

$$D = a^2 b\underline{D} \;.$$

Let us denote HL by

$$HL = a^2 b \;.$$

Then

$$D = HL\underline{D} \;.$$

This relation was called the similarity relation by Pogozhev. It should be mentioned that for the diffusion process the following equality holds:

$$u(\beta t) = \sqrt{\beta}\,u(t) \;.$$

Then

$$u(t) = a\underline{u}(bt) = \underline{u}(a^2 bt) = \underline{u}(HLt) \;,$$

where HL is the parameter of similarity and the relation of similarity has the form

$$u(t) = \underline{u}(HLt) = (HL)^{1/2}\,\underline{u}(t) \;. \tag{10.10}$$

Repeating previous calculations we can write the equations for densities:

$$\frac{\partial}{\partial t}\,\underline{p}(q,t) = -\sum_i \overline{\underline{\psi}_i}(\pi(t))\,\frac{\partial \underline{p}(q,t)}{\partial q_i} + \frac{1}{2}\sum_{i,j}\underline{D}_{ij}\,\frac{\partial^2 \underline{p}(q,t)}{\partial q_i \partial q_j}$$

$$\frac{\partial}{\partial t}\,p(q,t) = -\sum_i (HL)^{1/2}\,\overline{\underline{\psi}_i}(\pi(t))\,\frac{\partial p(q,t)}{\partial q_i} + \frac{1}{2}\sum_{i,j}\underline{D}_{ij}\,\frac{\partial^2 p(q,t)}{\partial q_i \partial q_j} \;.$$

This means that for the shift vector we have a transformation

$$\overline{\psi} = (HL)^{1/2}\,\underline{\psi} \;,$$

and the same results can be obtained for the non-stationary case.

10.5 Parametrization of the Model

Let us now return to the models in equations (10.2) and (10.3). We have to define an analytical form of the function $v(\underline{\alpha}, HL)$. Let $\eta(\tau)$ and $\underline{\eta}(\tau)$ be the number of interactions of particles during the period τ in the given and basal organisms respectively. These values are functionals of the trajectory of the interacting particle system

$$\underline{\eta}(\tau) = J(\underline{q}(t),\ 0 \le t \le \tau) = \underline{\mu}\tau$$
$$\eta(\tau) = J(q(t), 0 \le t \le \tau) = \mu\tau \ .$$

To explain this proposition we can consider the number of interactions as a random process of events which is stationary with intensity μ. In this case, the probability of z interactions is

$$P(z = k) = \frac{(\mu\tau)^k}{k!}e^{-\mu\tau} \ .$$

Then the average number of contacts is $\mu\tau$. Using the similarity relation we have

$$
\begin{aligned}
\mu\tau = \eta(\tau) &= J(u(t), 0 \le t \le \tau) \\
&= J(\underline{u}(tHL), 0 \le t \le \tau) \\
&= J(\underline{u}(s), 0 \le s \le \tau HL) = \underline{\mu}\tau HL
\end{aligned}
$$

or

$$\mu = \underline{\mu}HL \ . \tag{10.11}$$

The vector of α parameters in the model in equation (10.1) is proportional to the intensity of interactions, μ, and $\alpha = \mu\gamma$, where γ is a constant vector. From equation (10.11), for a given organism, we have

$$\alpha = \underline{\alpha}HL \ .$$

Thus,

$$v(\underline{\alpha}, HL) = HL\underline{\alpha} \ .$$

10.6 Similarity of the Homeostasis Levels

The reaction of the homeostatic system to perturbations is described by the linear system

$$\frac{d}{dt}x_t = Ax_t + B, \quad x_0 = c \ . \tag{10.12}$$

Here, c is a vector of disturbances, A is a matrix of parameters, and B is a vector which is interpreted as the rate of influx of particles into the interaction zone. Therefore, this vector is proportional to the shift vector ψ.

This allows us to consider the vector B as a lineal functional of ψ, and we can write

$$B = G(\overline{\psi}) = G\left((HL)^{1/2}\,\underline{\overline{\psi}}\right) = (HL)^{1/2}G(\underline{\overline{\psi}}) = (HL)^{1/2}\underline{B} \ .$$

According to the model in equation (10.12), the homeostasis level is

$$x^\infty = A^{-1}B \ .$$

Then, taking into account our previous results, for the individual organism we have

$$\frac{d}{dt}x_t = HL\underline{A}x_t + (HL)^{1/2}\underline{B}$$

where $\underline{A}$ and $\underline{B}$ are the parameters of the basal organism. So, we obtain

$$x^\infty = \frac{1}{HL}\,\underline{A}^{-1}\,\underline{B}(HL)^{1/2} = \underline{x}^\infty\,(HL)^{1/2} \ .$$

These transformations of parameters are sufficient for this work. The reader can find others in Pogozhev (1989). For example, for the effective volumes of interactions we have

$$W = \underline{W}\,HL^{3/2} \ .$$

10.7   The Model of Carbohydrate Metabolism

To confirm our theory, consider the model of carbohydrate metabolism which is known as the Bolier model (Antomonov *et al.*, 1971):

$$\begin{aligned}
\frac{d}{dt}y(t) &= -\alpha_1 y(t) \\
W\frac{d}{dt}x_1(t) &= \alpha_2 y(t) - \alpha_3 x_1(t)W - \alpha_4 x_2(t)W \\
W\frac{d}{dt}x_2(t) &= \alpha_5 x_1(t)W - \alpha_6 x_2(t)W \\
y(0) &= c, \ \ x_1(0) = x_2(0) = 0 \ ,
\end{aligned}$$

where y_t is the quantity of glucose in the intestines, $x_1(t) = G_t - G^*$, and G_t is the concentration of glucose in the blood. Also, $x_2(t) = I_t - I_t^*$,

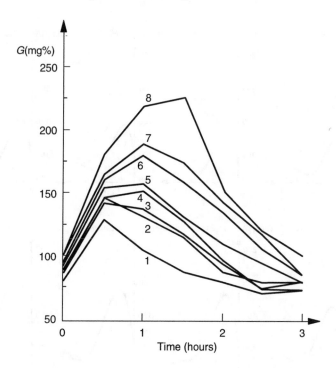

Figure 10.1. Sugar curves: G-blood sugar. Age groups: 1 = up to 10; 2 = 10–20; 3 = 20–30; 4= 30–40; 5 = 40–50; 6 = 50–60; 7 = 60–70; 8 = more than 70.

where I_t is the concentration of insulin in the blood, and G^* and I^* are related homeostasis levels. Let $\underline{\alpha}^i$, $i = 1, 2, \ldots, 6$, $\underline{G}^*$ and $\underline{I}^*$ correspond to the basal organism. Using the previous results for a given organism, we have

$$\frac{d}{dt}y(t) = -HL\underline{\alpha}_1 y(t)$$
$$\frac{d}{dt}x_1(t) = HL^{1/2}\underline{\alpha}_2 y(t) - [\underline{\alpha}_3 x_1(t) - \underline{\alpha}_4 x_2(t)]\,HL$$
$$\frac{d}{dt}x_2(t) = HL[\underline{\alpha}_5 x_1(t) - \underline{\alpha}_6 x_2(t)]$$
$$y(0) = c, \quad x_1(0) = x_2(0) = 0$$
$$x_1(t) = G_t - \underline{G}^*(HL)^{-1/2}$$
$$x(t)^* = I_t - \underline{I}^*(HL)^{-1/2}$$
$$W = \underline{W}\,HL^{3/2} \ . \tag{10.13}$$

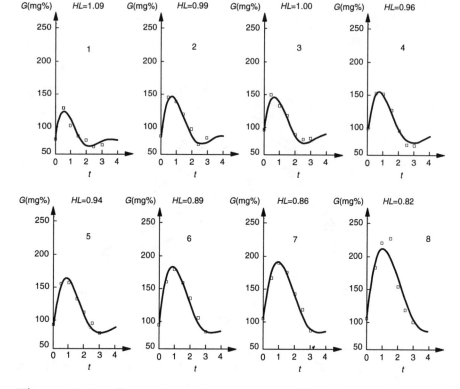

Figure 10.2. Comparison of sugar curves. The chart numbers and values of HL correspond to the age groups in *Figure 10.1*.

In the model the parameter HL takes into account the individual character of the metabolism process in a given organism. For the group of m patients we have H^1, HL^2, ..., HL^m individual parameters.

Here, we discuss some practical examples, following the work of Pogozhev. In *Figure 10.1* the dynamics of sugar in the blood are presented (Dilman, 1983). Every sugar curve corresponds to a single age group of healthy people and describes the average sugar dynamics in this group after glucose loading. These curves can be considered as being obtained from eight different patients with

$$X_m = \{x_t^i, \ t \in \Theta, \ i = 1, 2, \ldots, 8\} \ .$$

For X_m we have the model in equation (10.13) and the parameter HL takes into account the peculiarity of every curve. In our example, as a basis for the calculation of HL we consider the group of patients from

20 to 30 years old as a basal trajectory. Using this curve we can estimate $\underline{G}_\infty$, $\underline{I}_\infty$, and $\underline{\alpha}$. Now, for our example, we recalculate the parameters of the model for each curve using the relationships

$$\alpha = \underline{\alpha} HL$$
$$G_\infty = \underline{G}^*(HL)^{-1/2}$$
$$I_\infty = \underline{I}^*(HL)^{-1/2} \ .$$

Here, HL is an unknown parameter. For the group of patients from 20 to 30 years old, $HL = 1$. The results are presented in *Figure 10.2*.

From the graphs it can be seen that the distinctions between curves are completely described by the parameter HL. All these results show that the approach which is proposed here for parametrization of the model do not contradict the observed data. Of course, this approach should be tested with other models and data.

Appendix B

Generalized Index Construction

Let us denote the clinical state vector by $x = (x^1, x^2, \ldots, x^n)$, each component of which is observed in a clinic. Remembering that s is patient state severity according to the classification from Section 6.2 ($s = 0, 1, \ldots, r$), we consider x as an n-dimensional random value that is connected with the discrete random value s.

So we will let $E(x^i|s) = g^i(s)$, where $g^i(s)$ are known functions, $i = 1, 2, \ldots, n$, and E is the mathematical expectation operator. The functions $g^i(s)$ are regressions of x^i on s and show the average picture of changing x^i values and their dependence on s. Assume that indices are chosen in such a way that $g^i(s) \neq$ constant at any s. Choose the function $\phi(x)$ from the condition $E(\phi(x) - s)^2 \to \min$.

Write this expression in more detail, i.e., in the form of the complete mathematical expectation

$$E\left(\phi(x) - s\right)^2 = \sum_s p_s \, E\left[(\phi(x) - s)^2|s\right] \quad, \tag{B.1}$$

where p_s is the probability of the state with number s.

It is well known (Cramer, 1946) that at $E(\phi(x)|s) = s$ the expression reaches its minimum.

B.1 Linear Case

Let $g^i(s)$, $i = 1, 2, \ldots, n$, be linear and $g^i(s) = s$. In this case we can choose the linear form

$$\phi(x) = \sum_{i=1}^{n} \alpha^i x^i , \quad \sum_i \alpha^i = 1 .$$

Indeed,

$$E\left[\phi(x)|s\right] = E\left(\sum_i \alpha^i x^i | s\right) = \sum_i \alpha^i E(x^i|s) = s .$$

The coefficient of the generalized index α^i, $i = 1, 2, \ldots, n$, is defined by the minimum condition of expression (B.1) on the set $\{\alpha : \Sigma \alpha^i = 1\}$. We then construct the Lagrange function:

$$L(\alpha, \lambda) = \sum_s p_s E\left[\left(\sum_i \alpha^i(x^i - s)\right)^2 | s\right] + \lambda\left(\sum_i \alpha^i - 1\right) .$$

Differentiating $L(\alpha, \lambda)$ by $\alpha^1, \alpha^2, \ldots, \alpha^n$, and λ, and making the derivatives equal to zero, we obtain the following system of linear algebraic equations with respect to unknown coefficients:

$$\sum_{i=1}^{n} \alpha^i \sum_s p_s E\left[(x^i - s)(x^k - s)|s\right] + \lambda = 0 , \quad k = 1, \ldots, n$$

$$\sum_{i=1}^{n} \alpha^i = 1 . \tag{B.2}$$

The values

$$R^{ij} = \sum p_s E\left[(x^i - s)(x^j - s)|s\right] = \operatorname{cov}(x^i, x^j)$$

$$i, j = 1, 2, \ldots, n ,$$

form the covariance matrix R of the n-dimensional random value x given by $x = (x^1, x^2, \ldots, x^n)$.

Consider the case when the matrix R is diagonal (i.e., $x^1 - s$, $x^2 - s, \ldots, x^n - s$ are uncorrelated). From the first n equations, we obtain

$$\alpha^i = -\lambda/D^i, \quad D^i = R^{ii}, \quad i = 1, 2, \ldots, n . \tag{B.3}$$

From the last equation in (B.2) and the expressions in equation (B.3) we find

$$\lambda = -\left(\sum_{i=1}^{n} \frac{1}{D^i}\right)^{-1} .$$

Now it is easy to obtain a formula for α^i:

$$\alpha^i = \left(D^i \sum_{k=1}^{n} \frac{1}{D^k} \right)^{-1} .$$

One can see that the weight coefficients α^i $(i = 1, 2, \ldots, n)$ in the linear form $\phi(x)$ are chosen to be inversely proportional to the variance. Define the variance of $\phi(x)$ as follows:

$$E\left[\phi(x) - s\right]^2 = E\left(\sum_{i=1}^{n} \alpha^i(x^i - s) \right)^2 = \sum_{i=1}^{n} (\alpha^i)^2 D^i$$
$$= \left(\sum_{i=1}^{n} \frac{1}{D^i} \right)^{-1} ,$$

that is, D^i terms are positive and $D^i < \infty$, $i = 1, 2, \ldots, n$. Then, when $n \to \infty$ we obtain

$$E\left[\phi(x) - s\right]^2 \to 0 .$$

Therefore, in the situation where there is an unlimited increase in the number of uncorrelated indices, the error state estimate can be as close to zero as desired. This is the rationality of the generalized index construction. Unfortunately, in real situations we do not always have linear regression $E(x^i|s) = s$. However, one can try to reduce the non-linear case as discussed above. In other words, we can try to transform x^i, $i = 1, 2, \ldots, n$, with the help of functions $y^i = f^i(x)$ in such a way that

$$E\left(f^i(x^i)|s \right) = s, \quad i = 1, 2, \ldots, n, \quad s = 0, 1, \ldots, r .$$

Then for the new variables $Y = (y^1, y^2, \ldots, y^n)$, $y^i = f^i(x^i)$, the construction of $\phi(Y)$ is reduced to the problem discussed. The functions $f^i(x)$, $i = 1, 2, \ldots, n$, can be chosen in different ways. We derive them using the polynomial form:

$$f^i(x) = c_{0i} + c_{1i}x + c_{2i}x^2 + \cdots + c_{ri}x^r .$$

For simplicity, from this point on the index i will be omitted as we address the i-index transformation.

B.2 Non-linear Case

We need to define the polynomial coefficients from the condition of linear regression, i.e., as the solution of the following linear algebraic system:

$$E\left[f(x)|s\right] = c_0 + c_1 E(x|s) + E(x^2|s) + \cdots + c_r E(x^r|s) = s$$

$$s = 0, 1, \ldots, r \ . \tag{B.4}$$

Suppose that for each index the system in equation (B.4) is derived and the polynomials $f^{(i)}(x)$ such that

$$E\left(f^i(x^i)|s\right) = s, \quad s = 0, 1, \ldots, r$$

are found. Then each case considered reduces to the previous one and the generalized index has a form

$$\phi(x) = \sum_{i=1}^{n} \alpha^i f^i(x^i)$$

where α^i are the solutions of (B.2) with the elements of matrix R in the form

$$R^{ij} = \text{cov}\left(f^i(x^i), f^j(x^j)\right) \ .$$

The GI can be constructed in different ways. We discuss here the variant of its construction that is simply realized on the computer and now used in practice. In the general case, $\phi(x)$ can be represented as a non-linear function of linear form (see Marchuk, 1984), that is,

$$\phi(x) = F\left(y(x)\right) \ ,$$

where

$$y(x) = \sum_{i=1}^{n} \alpha_i x_i \ .$$

To find $F(y)$ we use the expansion

$$F(y) = F(0) + F'(0)y(x) + \frac{1}{2} F''(0)y^2(x) + \cdots \ . \tag{B.5}$$

From the results of clinical observations on the group of m patients we have the sample values of indices $x^i = (x^{1i}, x^{2i}, \ldots, x^{ni})$, $i = 1, \ldots, m$, and corresponding to these values the estimates of state s^i, $i = 1, \ldots, m$. The generalized index is represented as the sum of orthonormalized polynomials:

$$\phi(x) = \omega_0 Q_0(y) + \omega_1 Q_1(y) + \cdots + \omega_r Q_r(y) \ , \tag{B.6}$$

where

$$y = \sum_{i=1}^{n} \alpha_i x^i \ .$$

The terms $\omega, \omega_1, \ldots, \omega_r$ denote the coefficients and $Q_k(y)$ are orthonormalized polynomials, that is,

$$Q_k(y) = q_{k0} + q_{k1}y + \cdots + q_{kk}y^k$$

$$\sum_{i=1}^{m} [Q_k(y_i)]^2 = 1$$

$$\sum_{i=1}^{m} Q_k(y_i)Q_j(y_i) = 0$$

$$y_k = \sum_{i=1}^{n} \alpha_i x^{ik}, \ \ k,j = 1,\ldots,r, \ \ k \neq j \ . \tag{B.7}$$

The representation of the generalized index in the form of (B.6) enables us to simply calculate the coefficients ω_i as

$$\omega_i = \sum_{j=1}^{m} s_j Q_i(y_j), \ \ i = 1,\ldots,r \ .$$

For the derivation of the polynomials we use the Forsight recurrent relation:

$$\begin{aligned} \lambda_k Q_k(y) &= y Q_{k-1}(y) - \gamma_k Q_{k-1}(y) - \beta_k Q_{k-2}(y) \\ Q_0(y) &= \frac{1}{\sqrt{m}} \\ Q_1(y) &= \frac{(y-Y)}{\sqrt{D}} \ , \end{aligned}$$

where

$$\begin{aligned} Y &= \frac{1}{m} \sum_{i=1}^{m} y_i \\ D &= \sum_{i=1}^{m} (y_i - Y)^2 \\ \gamma_k &= \sum_{i} y_i Q_{k-1}^2(y_i) \end{aligned}$$

$$\beta_k = \sum_i y_i Q_{k-1}(y_i) Q_{k-2}(y_i)$$

$$\lambda_k = \left(\sum_i (y_i Q_{k-1}(y_i) - \gamma_k Q_{k-1}(y_i) - \beta_k Q_{k-2}(y_i))^2 \right)^{1/2} .$$

Naturally, the question of polynomial degree arises. Beginning the derivation, for example, with $r = 1$, one needs to increase the polynomial degree until the fact that beginning with $r = l$, ω_i at $i \geq l$ is not significant. To test the hypothesis that $\omega_r = 0$, we use the fact that the expression

$$\frac{\omega_r^2}{\sigma_{m-r-1}^2} ,$$

where

$$\sigma_{m-r-1}^2 = \frac{R_r}{m-r-1}$$

$$R_r = \sum_{i=1}^{m} s_i^2 - \sum_{i=1}^{r} \omega_i^2 ,$$

according to Hudson (1970) has a Fisher distribution with the number of degrees of freedom equal to $1, m - r - 1$ for the hypothesis $\omega_r = 0$. So, by increasing the polynomial power and testing each time the hypothesis that $\omega_r = 0$, one can construct the GI with minimal polynomial power. The algorithm considered enables one to derive the expression for the generalized index of severity without solving the algebraic system of equations. Its computer realization is very simple.

The practical application of the GI formula is based on the following. For an individual patient, all the terms $x^1(t), x^2(t), \ldots, x^n(t)$ in the GI are known for the time $t \geq 0$. The computation yields the estimated degree of gravity expressed in a continuum form rather than the four-point system. The continuous values of this estimate are of considerable interest and are especially important in studying the dynamics of a disease. Obviously, this application of the generalized index (i.e., using a scalar value to replace a vector) essentially simplifies the analysis of the dynamics of the organism's recovery. In Chapter 6 the application of the GI to the solution of some practical problems is shown.

References for Part II

Ada, G.L., and Jones, P.D., 1986, *Current Topics in Microbiology and Immunology* **128**:1–54.

Antomonov, Yu.G., *et al.*, 1971, *Mathematical Theory of the System of Sugar in Blood*, Naukova Dumka, Kiev, Ukraine.

Asachenkov, A.L., Pogozhev, I.B., and Zuev, S.M., 1992, *Parameterization in the Mathematical Models of Immuno-Physiological Processes*, WP-92-39, International Institute for Applied Systems Analysis, Laxenburg, Austria.

Asconas, B.A., Mullbacher, A., and Aschman, R.B., 1982, *Immunology* **45**:79–84.

Bakhvalov, N.S., Zhidkov, N.P., and Kobelkov, G.M., 1987, *Numerical Inference*, Nauka, Moscow, Russia.

Balakrishnan, A.V., 1984, *Kalman Filtering Theory*, Springer-Verlag, Berlin, Heidelberg, New York.

Balakrishnan, A.V., 1986, A Note on the Marchuk–Zuev Identification Problem, in A.V. Balakrishnan, A.A. Dorodnitsyn, J.L. Lions, eds., *Vistas in Applied Mathematics*, Optimization Software, Inc., Publication Division, New York, NY, USA.

Bukhgolts, N.N., 1972, *Basic Course of Theoretical Mechanics*, Nauka, Moscow, Russia.

Cramer, H., 1946, *Mathematical Methods of Statistics*, Princeton University Press, Princeton, NJ, USA.

Dilman, V.M., 1983, *Four Models of Medicine*, Meditsina, St. Petersburg, Russia.

Gardiner, C.W., 1984, *Handbook of Stochastic Methods for Physics, Chemistry and the Natural Sciences*, Springer-Verlag, Berlin, Heidelberg, New York.

Hudson, D., 1970, *Statistics for Physicists*, Mir, Moscow, Russia.

Kalyaev, D.V., 1987, pp. 26–40 in A.A. Totolyan and R.Ya. Polyak, eds., *The Strategy of a Pathogen in the Host Organism* (in Russian), Institute of Experimental Medicine, St. Petersburg, Russia.

Kalyaev, D.V., and Zuev, S.M., 1991, Algorithms for Statistical Estimation of Coefficients of Ordinary Differential Equations Using Observational Data, *Soviet Journal of Numerical Analysis and Mathematical Modelling* **6**(1): 1–25.

Lancaster, P., 1969, *Theory of Matrices*, Academic Press, New York, NY, USA.

Libbert, E., 1982, *Principles of General Biology* (Russian translation from German 4th edn. published by Fischer, Stuttgart, Germany), Mir, Moscow, Russia.

Liptser, R.S., and Shiryaev, A.N., 1977, *Statistics of Random Processes*, Springer-Verlag, New York, NY, USA.

Luria S., *et al.*, 1978, *General Virology*, 3rd edn., John Wiley & Sons, New York, NY, USA.

Marchuk, G.I., 1981, *Methods of Computational Mathematics*, Springer-Verlag, Berlin, Heidelberg, New York.

Marchuk, G.I., 1983, *Mathematical Modelling in Immunology and Medicine*, North-Holland, Amsterdam, The Netherlands.

Marchuk, G.I., 1984, *Mathematical Models in Immunology* (English version), Optimization Software, Inc., Publication Division, distributed by Springer-Verlag, New York. (Russian version: 1980, Science Press, Moscow.)

Marchuk, G.I., Polyak, R.Ya., Zuev, S.M., and Kalyaev, D.V., 1988, Mathematical Modelling of Interaction in the Virus-Host System (Experimental Influenzal Infection), *Mendeleev Chemistry Journal* **33**(5), Allerton Press, Inc., New York, NY, USA.

Mudrov, V.I., and Kushko, V.L., 1983, *Inference of Measurement Processing*, Padio i svias, Moscow, Russia.

Nisevich, N.I., Marchuk, G.I., Zubikova, I.I., and Pogozhev, I.B., 1984, *Mathematical Modelling of Viral Diseases* Optimization Software, Inc., Publication Division, New York, NY, USA.

Paul, W.E., ed., 1984, *Fundamental Immunology*, Raven Press, New York, NY, USA.

Pogozhev, I.B., 1988, *Application of the Mathematical Models of Disease in Clinical Practices* (in Russian), Nauka, Moscow, Russia.

Pogozhev, I.B., 1989, *Intensity of Interactions in Liquid Media of an Organism*, Institute of Numerical Mathematics, USSR Academy of Sciences, Moscow, Russia.

Polyak, R.Ya., 1987, pp. 10–25 in A.A. Totolyan and R.Ya. Polyak, eds., *The Strategy of a Pathogen in the Host Organism* (in Russian), Institute of Experimental Medicine, St. Petersburg, Russia.

Rao, C.R., 1973, *Linear Statistical Inference and its Applications*, 2nd edn., John Wiley & Sons, New York, NY, USA.

Shmidt-Nielson, K., 1987, *Sizes of Animals: Why Are They So Important?*, Mir, Moscow, Russia.

Ventcel, A.D., and Freidlin, M.I., 1975, *Fluctuation in Dynamic Systems by Influence of Small Random Perturbations*, Nauka, Moscow, Russia.

Zhdanov, V.M., 1985, *Atlas of Viral Cytopathology* (in Russian), Meditsina, Moscow, Russia.

Zuev, S.M., 1986, Statistical Estimation of the Coefficients of Ordinary Differential Equations using Observational Data, *Soviet Journal of Numerical Analysis and Mathematical Modelling* **I**(3):235–244.

Zuev, S.M., 1988, *Statistical Estimation of Parameters of Mathematical Models of Disease* (in Russian), Nauka, Moscow, Russia.

Part III

Mathematical Models in Cancer Research

Introduction

At the present time about 100 different tumors are distinguishable in humans. Each of them is characterized by its rate of growth, by the age at which it occurs most often, the way it spreads to other tissues, and its mortality dynamics.

Cancer is an "uncontrollable" growth of the organism's cells which germinate posterity in neighboring organs and tissues. This process is called invasion and may be observed in the process of metastasis, in which the tumor's cells spread via lymph or blood. The spreading of disease by a metastatic process is a principal reason for patients' deaths. In this case, the tumor can not normally be removed by radical surgery and is inaccessible to radiation treatments. This failure of traditional cancer therapy is the motivation for further research on control therapy. The case for immunotherapy has been given encouragement by the survival of certain patients for whom the cancer has metastasized. Apparently, the immune system plays a strong role in such cases, and its therapeutic stimulation is studied here. Some tumors, for some unknown reason, do not metastasize: for example, some skin cancers. Such tumors can be treated easily if, of course, the region germinating them does not lead to patient death. On the other hand, normal bone marrow and lymph cells are distributed in the organism and it is not surprising that tumors from such cells (leucosis, etc.) spread throughout the organism from the beginning of the disease. In the majority of cases, tumor growth is somewhere between these two extreme cases.

In the last century, the classification of human tumors according to their localization, cell species, and the peculiarity of their growth was created by morphologists. Such classification is very important because in many cases each class has its own progression of development. Some of them lead to death very fast and others do not; most tumors arise in the elderly, but there are others which arise only in children; some develop only in inhabitants of industrial countries, others do not, etc.

Here we study the problems of mathematical modeling of tumor growth and the related data processing in some depth to understand the process in question and to formulate the optimal treatment policy. Two classes of mathematical models may be considered.

1. The class for which the tumor volume or the number of tumor cells can be calculated by direct measurements or by using specific "tumor markers" [carcinoembryonic antigen (CEA), α-foetoprotein, etc.].
2. The class for which there are no sufficiently specific "tumor markers".

The first class requires a direct measurement of tumor character-istics (lineal dimensions, volume or mass, number of tumor cells, etc.). Traditionally, accurate measurements can only be made on tumors grow-ing in a few sites. For this reason, the majority of studies have examined lung metastases using serial X-rays. Measurements of primary tumors in tissues other than the lung are rare. Another approach is to use "tu-mor markers" – substances produced by tumor cells, made radioactive, and then measured by their radioactivity. Modern scanning machines, while still somewhat expensive, do provide the possibility of more pre-cise estimates of tumor size to provide credence to these methods. Here we deal with so-called "tumor growth models". Traditional, solid tumor growth models are considered in Chapter 11. A different situation arises when the patient has several new growths at the same time which have different morphological structure and size. Here, it is more convenient to use other characteristics of tumor growth. One realistic example is considered in Chapter 12.

In the majority of cases it is impossible to measure the volume of a tumor and the number of tumor cells when there are no specific tumor markers. To study the tumor growth process in such cases we may use clinically measured laboratory indices. This calls for the development of more sophisticated models such as those studied here.

Let $z(t) \geq 0$ and $z : \mathbb{R} \to \mathbb{R}$ be a function describing the time course of the volume or number of tumor cells. Suppose that the physiolog-ical or immunological indices [molecular concentrations (such as those of antibodies, antigens, appropriate lymphokines, or other molecular substances) and cellular populations (which might refer to B, T, and macrophage cell lineages, etc.)] at time t are given by the vector $x(t) = (x_1(t), \ldots, x_n(t))^T$, where $x_i \geq 0$, $i = 1, \ldots, n$. There are two ways to construct the equations for the model. The first is based on the following hypothesis.

Hypothesis III.1
Tumor cells differ antigenically from normal cells, and the host defense mechanisms are capable of recognizing and exploiting these differences.

Here, we deal with mathematical models based on the cellular and molecular kinetics studied in Chapter 2. This process was determined using conservation equations and chemical mass–action principles. Good examples are given by Rescigno and DeLisi (1977), Grossman and Berke (1980), Lefever and Garay (1978), De Boer and Hogeweg (1986), and Mohler and Lee (1990).

One of the first models describing tumor growth in an organism was proposed by Rescigno and DeLisi in 1977. The tumor growth is considered as two interacting populations: lymphocytes, $x(t)$, and tumor cells, $z(t)$. Assuming that only the cells situated on the surface of a tumor are available to be killed by lymphocytes, the equation of the model has the form

$$\frac{dz(t)}{dt} = \lambda z(t) - \alpha(.)x(t)z^{2/3}(t)$$

$$\frac{dx(t)}{dt} = \beta(.)x(t)z^{2/3}(t) - \mu\left(x(t) - q\right)$$

$$\alpha(.) = \frac{\alpha}{1 + x(t)}, \quad \beta(.) = \frac{k\left(1 - x(t)/x_{\max}\right)}{1 + x(t)} \ .$$

If we suppose that all tumor cells are available to the lymphocyte, then

$$\frac{dz(t)}{dt} = \lambda z(t) - \alpha(.)x(t)z(t)$$

$$\frac{dx(t)}{dt} = \beta(.)x(t)z(t) - \mu\left(x(t) - q\right) \ .$$

In spite of such a simple explanation of the anti-tumor reaction (only one cytotoxic mechanism was studied), the model solutions are in close agreement with experimental facts and intuition. For example, if we inject a small number of tumor cells into an organism, then the tumor is destroyed, but if the initial number of tumor cells is above some threshold value then they increase in number in spite of the fact that the immune reaction has begun fairly early. However, in the framework of this model it is impossible to describe the tumor development *de novo*. Particular mechanisms can not be investigated. Thus, a more realistic model is studied in Chapter 13. Here, the presentation follows the work of Mohler and Lee (1990).

Unfortunately, in many cases, in practice, tumor cells are weakly antigenic and, consequently, we have a weak immune response as shown in the model in Chapter 13. Nevertheless, we can formulate the following hypothesis.

Hypothesis III.2

Tumor growth leads to disorder in the normal functioning of the main homeostatic systems of the organism, which is manifested by the deviations of measurable indices from the values that correspond to the healthy state of the organism (no tumor).

Using these hypotheses we can study the dynamics of the measurable indices $x(t) \in \mathbb{R}^n$ (Asachenkov, 1990). Then, the monitoring of the body state based on the analysis of the available information is a necessary prerequisite for the right choice of treatment. In Chapter 14, we study such a model in detail (Marchuk et al., 1989).

Anti-cancer drugs do influence tumor growth, but the real problem is to describe the therapy (control) in mathematical terms. Another problem is the selection of a suitable performance criterion which expresses the desired improvement in the health of an individual in quantitative (mathematical) terms.

In cancer control research the following traditional cost functional may be used:

$$J(u) = q\left(z(T)\right) + \int_0^T R(z, u)\, dt\ ,$$

where $q\left(z(T)\right)$ is associated with still-viable cancer cells in the patient at the end of the period of therapy, and the second term estimates the cumulative drug toxicity over the time period $[0, T]$. The optimization problem is to determine the control (administration of drugs) required so as to minimize the performance index.

Also, there is another method. Let $y_r(t)$ be a known function which prescribes the "desired" time course for a measured variable and which reflects the experience of the investigator. Define $e(t, u) = x(t, u) - y_r(t)$, where $e(t, u)$ is the deviation between the model-derived value $x(t, u)$ and its desired level. It seems reasonable to use an anti-cancer drug in such a manner, avoiding life-threatening toxicities, that the measurable variables follow the curve given by $y_r(t)$ with the minimal toxicity effect (Swan, 1984) so that

$$J(u) = \int_0^T \left(e^T(t, u)Q(t)e(t, u) + u^T(t)Ru(t) \right) \, dt \to \min_{u \in U} \ ,$$

where Q and R are appropriate matrices.

There are some questions, of course, with respect to such performance indices. The function $y_r(t)$, as a rule, is determined through examination by experts. Therefore, it contains some elements of subjectivity. Is it possible to choose this function using only objective information? This cost function comes from technical applications: there are no obvious reasons why it will be useful for cancer therapy. We will discuss this problem in Chapter 14. Last but not least, where is our patient? One approach to individual management, based on the available information, and future directions are discussed in Chapter 15.

Chapter 11

The Tumor Growth Models

Here, we consider traditional tumor growth models. The most detailed information comes from experimental studies of animal tumors. Clinical data are more difficult to obtain and show a greater degree of inter-tumor variation. As a rule, tumor growth is described by the number of tumor cells or volume or tumor mass. However, in some cases the patient has several new growths at the same time, which have different morphological structures and sizes, and it may be more convenient to use other characteristics of tumor growth. One realistic example is given in Chapter 12.

The following is a useful definition of cancer.

> Cancer is not a well defined disease. It is the name applied to a group of conditions in which inappropriate cell proliferation leads to an excess of cells causing disruption of normal tissue architecture. In general, such a cellular excess is called *neoplasia*, and the clinical conditions to which it gives rise are known collectively as *neoplastic disease*. Not all instances of neoplasia are described as cancer; this name is reserved for those which possess additional properties. A common wart, for example, is an instance of neoplastic disease – a localized excess of skin cells – but it rarely constitutes a serious clinical condition. This is because the tumor is self-limiting in growth and its constituent cells usually tend to stay in the same place, such tumors are said to be *benign*. By contrast, *malignant* tumors, those considered to be examples of cancer, seem to have no built-in limitation to growth and, most importantly, their cells are prone to spread or *metastasize* to other parts of the body. [Wheldon, 1988]

Most, but not all, types of cancer result from the transformation of a single stem cell. The transformation affects the genes of the cell, changing the structure of the DNA or pattern of gene expression. As a result, changes are made in the homeostatic regulation of the number of cells and a cell continues to generate progeny when it is biologically inappropriate to do so. The progeny are the daughters of the transformed cell which inherit the malignant property. It should be mentioned that the cells in most tumors are quite heterogeneous with respect to their properties, which suggests that cancer cells continually modify their properties during the growth of the tumor.

Pathologists classify a tumor with respect to the origin of its tissue. Thus an adenocarcinoma arises from glandular epithelium (*adeno* = gland, *carcinoma* = tumor in tissue of epidermal origin) and an osteosarcoma arises from bone (*osteo* = bone, *sarcoma* = tumor in tissue of mesenchymal origin). The word *anaplastic* is used to describe a tumor with little or no evidence of differentiation (Hill and Tannock, 1987).

11.1   Measurement of Tumor Size

Unfortunately, tumors can be detected only if they reach macroscopic dimensions. Typically, this means that a cell population of around 10^9 cells will be present, at least 30 generations since the first post-malignant division of the original, transformed stem cell. The fact that detection is biologically late has important consequences for the subsequent history of the disease (Wheldon, 1988). This situation is illustrated in *Figure 11.1*. (The illustration assumes single-cell origin, detectability at 10^9 cells, and death at $10^{12} - 10^{13}$ cells, with exponential growth kinetics throughout. The actual time spent in the visible phase is probably somewhat underestimated because tumor growth is not usually exponential throughout, but slows down with increasing size.)

11.1.1   Measurement of tumors in animals

The methods for measuring tumor size can be divided into measurements of linear dimensions, volume (or mass), and the number of cells. These methods have been reviewed by Steel (1977), whose terminology we follow.

Usually, two perpendicular dimensions (length and breadth) are easy to measure, but with tumors that attach themselves to and invade the

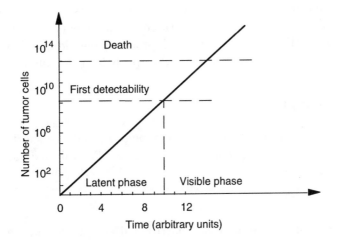

Figure 11.1. An illustration of the proportion of the life-history of a tumor which is spent in the visible and latent phases.

body wall, the third dimension is technically more difficult. A similar situation often arises with radiographically detected tumors in the human lung, where the third dimension is sometimes unobtainable, even with a lateral projection. The diameter of a subcutaneous tumor is measured with vernier calipers. The volume of the tumor may then be calculated from an assumed model of its shape. For deep-seated tumors, the volume is often estimated from X-ray radiography or imaging modalities such as ultrasonography, radionucleide imaging, and CAT or MRI scanning. The volume of the tumor is a more informative measure than a single, linear dimension, and its usefulness is unaffected by changes in geometrical form. Direct, *in situ* measurement of a tumor's volume is a difficult procedure.

Tumors usually grow either as nodules, such as lung metastases, or as sheets, such as superficial, spreading, malignant melanoma. Most nodules are approximately ellipsoid with one major axis of length L and two minor axes of lengths W and H, respectively, which are mutually perpendicular. Let the number of cells of a hematological tumor, or the total volume (cells and stroma) of a solid tumor, be represented by the symbol z, then the estimated volume of such a nodule is given by the formula

$$z = (\pi/6)LWH .$$

The majority of ellipsoid nodules have volumes which can be generated by revolution of an ellipse, with one major axis (L) and two, equal, minor axes ($W = H$), also termed prolate spheroid. These nodules, which have the shape of an American football, have volumes given by the formula

$$z = (\pi/6)LW^2 \ .$$

Sheets are usually elliptical, with one long (L) and one short (W) axis and a depth (H) which may be quite small. The approximate volume of such a sheet is

$$z = (\pi/4)LWH \ .$$

Most solid tumors may be "partitioned" or divided into a mixture of sheets and nodules. The total volume can then be calculated by summing the volumes of the component parts. In the majority of cases, especially for experimental tumors, the volume can be calculated within an acceptable degree of accuracy by treating the tumor as a single spheroid (Norton and Simon, 1979).

The alternative approach to obtaining the volume of a tumor from linear dimensions is the calibration curve method. An example of such a curve is shown in *Figure 11.2*. Each point represents measurements on a single, transplanted rat tumor. The area of the tumor is simply the product of two superficial diameters (measured with calipers), and its weight is a defect measurement made immediately after excision. In *Figure 11.2(b)* the data are plotted using double logarithmic coordinates in order to demonstrate the departure of the points from a 2/3 power relationship (from Steel *et al.*, 1977). It should be mentioned that there are no reliable methods for measuring tumors less than a few millimeters. Tumors in experiments on animals have been studied by measuring physical dimensions followed by surgical excision and weighing.

11.1.2   Measurement of human tumors

Estimates of the growth rates of untreated human tumors have been limited by the following constraints (Tannok, 1987):

- Only tumors that tend to be unresponsive to therapy can ethically be followed without treatment, although some data are available from older studies on the growth of tumors such as lymphoma, which are now treated aggressively with drugs.

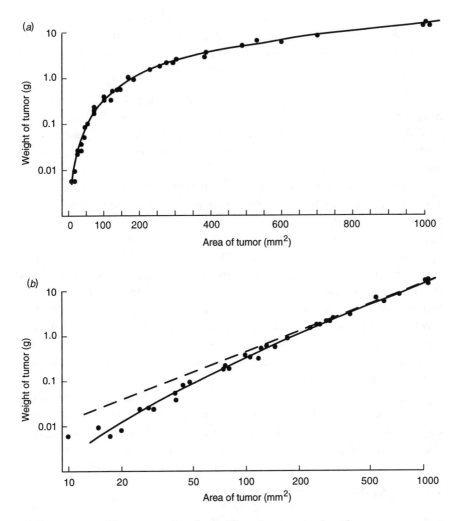

Figure 11.2. An example of a calibration curve for the measurement of the size of an *in situ* experimental tumor: (*a*) shows the weight of the tumor plotted against its area; (*b*) shows the same data plotted using double logarithmic coordinates.

• Accurate measurements can only be made on tumors growing in a few sites. For this reason, the majority of studies have examined lung metastases using serial X-rays. Measurements of primary tumors in tissues other than the lung have been rare, and are probably subject to substantial errors of measurement.

- The period of observation is restricted to that between the time of detection of the tumor and either the death of the host or the initiation of some form of therapy; this time interval is only a small fraction of the history of the tumor's growth.

The practicalities of in vivo measurement of the size of tumors in humans depends on the site of the tumor. Superficial lesions may be measured with calipers. Deeper lesions may be measured once at the time of an operative procedure. Useful measurements on tumors in humans could be made using radiographic techniques. The resolution of tumors detected by X-rays has been increased by xeroradiography and ionography and by computerized techniques of image processing. It has been shown that ultrasonic scanning gives good resolution of some tumors within the abdomen.

An alternative approach uses "tumor markers" – substances produced by tumor cells, made radioactive, and then measurements of the radioactivity are taken. Proteins whose production is specific to tumor cells have been found in the plasmacytomas of mice, human myeloma, and choriocarcinoma. Carcinoembryonic antigen (CEA) and α-foetoprotein are also indicators of the extent of certain neoplastic diseases, but they are seldom sufficiently specific to give an accurate indication of tumor size (Steel, 1977).

In many cases, the volume of the tumor is not the most biologically meaningful index of tumor growth. The number of cells, rather than volume, is the most meaningful parameter. The level of the biochemical (tumor) marker provides a reasonable estimate of the number of cells.

11.2   General Remarks

Tumor growth models have the following form of cellular conservation, as formulated in Chapter 2:

$$\frac{dz}{dt} = \text{source rate} + \text{growth rate} - \text{loss rate} \; , \tag{11.1}$$

where $z(t)$ is a size characteristic of the tumor, for example, volume or number of tumor cells. In the absence of external control or treatment, equation (11.1) can be written in the following general form:

$$\frac{dz(t)}{dt} = u(z)z(t) \equiv f(z)$$

$$z(0) = z^0 > 0, \quad t \geq 0, \quad z \in \mathbb{R}^1 . \tag{11.2}$$

Here, as a rule, $u(t)$ is a non-linear function of $z(t)$. This equation is a conservation equation for the population of tumor cells.

The solution of equation (11.2), as a rule, is a monotonic increasing function with saturation. It is easy to show that equation (11.2) has no periodic solution. Following the method used by Murray (1989), suppose that equation (11.2) has a periodic solution with period T: that is, $z(t + T) = z(T)$. Multiply the equation by dz/dt and integrate from t to $t + T$ to get

$$\int_t^{t+T} \left(\frac{dz}{dt}\right)^2 dt = \int_t^{t+T} f(z)\frac{dz}{dt}dt = \int_{z(t)}^{z(t+T)} f(z)dz = 0 ,$$

since $z(t + T) = z(T)$. However, the left-hand integral is positive, since $(dz/dt)^2$ cannot be equal to zero. So we have a contradiction, and the single scalar equation cannot have periodic solutions.

It should be mentioned that a population of cells growing by cellular fission may be modeled as a stochastic process, as noted in Chapter 2. The same is true throughout nature. Still, the deterministic models discussed here are useful in that they describe average populations.

From Taylor's theorem, equation (11.2) may be expanded as a power series:

$$\frac{d}{dt}z = a + bz + cz^2 + dz^3 + \cdots .$$

According to Burnet's theory, some of the normal cells continuously transform into malignant cells at a very small rate, denoted by $a > 0$. Unfortunately, this parameter cannot be measured directly. Indirect estimates are given by, for example, Garay and Lefever (1987). For the multitude myeloid growth, the probability that a plasmablast transforms and avoids immune surveillance is $10^{-16} - 10^{-17}$. Assuming that all cells which transform become tumor cells, we can estimate the intensity of carcinogenesis. For ideal tissue with 10^6 cells/mm^3 and a generation time of 10 days the intensity of carcinogenesis is about $10^{-11} - 10^{-12}$ tumor cells/day/mm^3. So, in many cases the term $a > 0$ may be omitted and we may write

$$\frac{d}{dt}z = bz + cz^2 + dz^3 + \cdots .$$

For the sake of simplicity we can now consider just one or two terms or more combinations of terms.

Example 11.1: Exponential model
Consider the linear equation

$$\frac{dz}{dt} = u(t)z \ , \quad z(0) = z^0 > 0 \ , \quad t \geq 0 \ ,$$

where z is the number of tumor cells and $u(t)$ is a specified rate of growth. The solution of this equation has the form

$$z(t) = z^0 \exp\left(\int_0^t u(\tau)d\tau\right) \ .$$

Assume that $u(t) = \gamma \equiv$ constant > 0, then we have a classical result

$$z(t) = z^0 \exp(\gamma t) \ .$$

Sometimes this equation is used for modeling the initial growth period when the effect of saturation can be neglected. The question of how tumors really grow during "latency" is an important one. Experimental data support the exponential model of growth during latency.

The experimental evidence comes from studies of the relationship between the number of tumor cells implanted in an animal used in the experiment and the time for the resultant tumor to become detectable (the latent period). The important characteristic of tumor growth is the time taken for the volume to double, given by $T_D = \ln 2/\gamma = 0.693/\gamma$. The doubling times for some human tumors are given in *Table 11.1* [from data reviewed by Steel (1977)]. In 1798 this equation was used by Malthus for describing population growth, but was actually suggested earlier by Euler.

Example 11.2: Gompertz growth
If in equation (11.2) we put $u(t) = Ae^{-Bt}$, where $B > 0$, then

$$z(t) = z^0 \exp\left(\int_0^T Ae^{-B\tau}d\tau\right) = z^0 \exp\left(\frac{A}{B}\left(1 - e^{-Bt}\right)\right) \ .$$

This equation has the form of the Gompertz function. The asymptotic limit proves to have some interesting properties. However, it should not be regarded as a physically meaningful number which applies to the tumor in practice. In most cases $N\infty = z^0 \exp(A/B)$ turns out to be a number so large that the tumor burden would result in the death of the host long before it was reached.

Table 11.1. Volume-doubling time (T_D) for representative human tumors.

Tumor type	Number of tumors	Volume-doubling time in weeks (T_D) (geometric mean value)
Primary lung cancer		
Adenocarcinoma	64	21
Squamous–cell carcinoma	85	12
Anaplastic carcinoma	55	11
Breast cancer		
Primary	17	14
Lung metastases	44	11
Soft-tissue metastases	66	3
Colorectal cancer		
Primary	19	90
Lung metastases	56	14
Lymphoma		
Lymph node lesions	27	4
Lung metastases of:		
Carcinoma of testis	80	4
Childhood tumors	47	4
Adult sarcomas	58	7

The asymptotic limit is best regarded as a mathematical abstraction which gives the right sort of curvature in the later stages of tumor growth, but which may not be physically achievable or even approachable (see Wheldon, 1988).

In some cases, the growth of a solid tumor seems to follow the Gompertz curve very precisely, at least in the early stages. Some examples are given in *Figure 11.3* and *Table 11.2*. The Gompertz growth function is completely empirical and there is no obvious reason why tumors should grow according to this function (Gyllenberg and Webb, 1989). It is intuitively unreasonable that the specific growth rate $u(t)$ should be a function of clock-time. More likely, $u(t)$ depends on some other set of biological variables and the time dependence arises indirectly. The situation in which the function $u(t)$ is a function of observed variables is studied in Chapter 14.

Gompertz published his law of human mortality in 1825. One hundred years later Wright (1926) proposed the use of this equation to describe the growth of individual organisms. However, the Gompertz

Table 11.2. Calculated parameters for tumor growth curves in animals (from Steel, 1977). (For an explanation of terms in the table heading see footnote a.)

Name	V^0 cm^3	A days^{-1}	B days^{-1}	V_{max} cm^3	SD_d days	t_0 days
(1) KHJJ sarcoma-like mouse (est.[b])	2.66×10^{-5}	1.386	0.116	3.99	0.86	4.7
(2) EMT6 sarcoma: mouse (est., adapted)	4.04×10^{-6}	1.050	0.0801	1.49	1.25	8.0
(3) KHT sarcoma: mouse (est.)	1.03×10^{-4}	0.997	0.0787	32.6	1.27	7.7
(4) Ca755 adenocarcinoma: mouse (est.)	9.6×10^{-5}	2.17	0.200	4.95	0.5	2.3
(5) Erlich carcinoma: mouse (est., solid)	2.26×10^{-2}	0.456	0.102	1.94	0.98	−4.3
(6) Erlich carcinoma: mouse (est., solid)	1.28×10^{-3}	1.14	0.129	8.81	0.78	−1.9
(7) Lewis Lang carcinoma: mouse (est.)	2.92×10^{-2}	0.526	0.0849	14.3	1.18	−1.28
(8) Sa180 sarcoma: mouse (est.)	1.17×10^{-3}	2.15	0.258	4.87	0.5	0.73
(9) NCTC2472 fibrosarcoma: mouse (est.)	1.26×10^{-3}	0.809	0.0916	8.69	1.10	2.7
(10) B16 Melanoma: mouse (est.)	7.10×10^{-4}	0.914	0.106	3.95	0.95	−2.1
(11) C3H carcinoma: mouse (spontaneous)	3.76×10^{-2}	0.177	0.0311	11.2	3.22	−6.2
(12) C3H carcinoma: mouse (spontaneous)	4.79×10^{-2}	0.144	0.0270	10.8	3.76	−9.1
(13) C3H carcinoma: mouse (1X[c])	2.54×10^{-3}	0.32	0.0321	54.6	3.12	11.4
(14) C3H carcinoma: mouse (est.)	3.11×10^{-5}	1.575	0.110	48.4	0.908	6.5
(15) C3H carcinoma: mouse	3.76×10^{-5}	0.369	0.0320	3.81	3.13	16.0
(16) PLa1 plasmacytomas: hamster (est.)	1.84×10^{-2}	0.789	0.107	29.3	0.938	0.61
(17) MM1 melanotic melanoma: hamster (est.)	3.80×10^{-3}	0.530	0.0554	44.0	1.81	5.9
(18) Walker carcin256: rat (est.)	4.96×10^{-6}	3.47	0.2048	113	0.5	4.4
(19) BA1112 rhabdomyosarcoma: rat (est.)	1.54×10^{-2}	0.254	0.0263	242	3.81	12.7
(20) BICR/A1 mammary: rat (est.)	1.02×10^{-5}	1.49	0.0966	50.3	1.24	8.3

Table 11.2. Continued.

Name	V^0 cm^3	A days^{-1}	B days^{-1}	V_{max} cm^3	SD_d days	t_0 days
(21) BICR/A2 fibrosarcoma: rat (21X)	0.217	0.163	0.0353	21.8	2.84	−11.4
(22) BICR/M1 sarcoma: rat (est.)	1.12×10^{-2}	0.950	0.1053	93.3	0.953	2.5
(23) BICR/M2 fibrosarcoma: rat (29X)	0.124	0.061	0.0060	14.0	–	–
(24) BICR/a3 osteogenic sarcoma: rat (4X)	1.89×10^{-4}	0.915	0.676	32.3	1.32	7.3
(25) BICR/A4 adenocarcinoma: rat (4X)	2.34×10^{-2}	0.146	0.0073	–	5.7	–
(26) BICR/A8 adenofibroma rat (2X)	1.06×10^{-3}	0.073	0.00545	696	18.4	121
(27) ^{90}Sr-induced osteosarcoma: mouse (primary)	9.18×10^{-4}	0.288	0.0345	3.91	2.91	5.5
(28) Spontaneous and induced adenosarcoma: rat (primary)	0.418	0.0846	0.0145	141	6.91	−11.7
(29) DMBA-induced adenosarcomas: rat (primary)	0.310	0.0922	0.0210	25.01	4.78	−21.6
(30) Radiation-induced mammary: rat (primary)	0.0440	0.0905	0.0153	16.30	6.56	−10.1
(31) Virus-induced skin carcinoma: rabbit (transplanted)	6.57×10^{-5}	1.390	0.0949	150.9	1.06	7.9
(32) Lang metastases of melanoma: dog metastases	6.50×10^{-2}	0.091	0.0168	14.63	5.97	−14.5
(33) Osteosarcoma:						
(a) dog metastases	0.648	0.0206	0.00814	8.14	12.32	−123
(b) dog metastases	5.91	0.0488	0.0214	57.8	4.69	−51.8
(c) dog primary	222.3	0.0138	0.0093	988	10.8	−166
(34) Venereal tumor	0.291	0.346	0.0792	22.9	1.27	−5.8

[a] The terms V_0, A, and B are the parameters of the Gompertz curve fitted to the published growth data; SD_d is the tumor-doubling time at a volume of one-thousandth of V_{max}; t_0 is the position of the time-zero of the standard curve on the originally published time-scale; V_{max} was more than 1000 times larger than tumors at the midpoint of the growth curve (SD_d is the doubling time at this mid-point).

[b] The term "est." indicates a transplanted tumor which undergoes frequent passages.

[c] X and a number indicates the number of passages.

Source: Steel (1977).

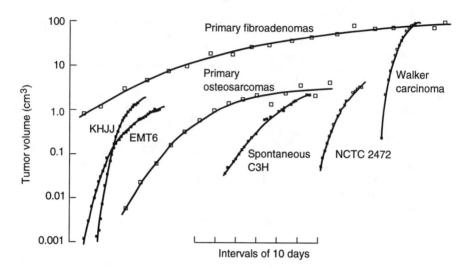

Figure 11.3. Growth curves for seven experimental tumors whose growth has been studied in detail. In each case the data fit a Gompertz equation and the corresponding parameters are listed in *Table 11.2*.

model owes its present popularity to Laird (1964) who developed its use as a growth model in a more systematic way.

11.3 Other Models

If the growth rate depends on the population at an earlier time, $t - T$, rather than that at time t, we can use differential equations with a delay factor. A simple example is an extension of the logistic equation

$$\frac{dz}{dt} = z(t)\left[\lambda_1 - \lambda_2 z(t - T)\right] \ , \quad z(t) = g(t) \ , \quad -T \le t \le 0 \ .$$

This equation is a model for the delay effect: for example, the time taken to reach maturity. A more accurate model is the convolution-type model:

$$\frac{dz}{dt} = z(t)\left(1 - K^{-1}\int_{-\infty}^{t} w(t - s)z(s)\,ds\right) \ .$$

Here, $w(t)$ is a weighting factor which indicates how much emphasis should be given to the size of the population at earlier times in order to determine the present effect on resource availability. A typical form of

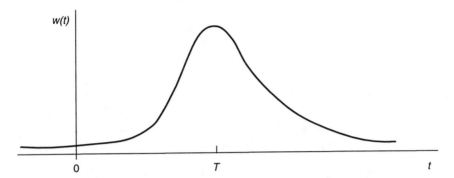

Figure 11.4. Typical weighting function for an integrated delay effect on growth limitation in the delay model.

$w(t)$ is given in *Figure 11.4*. In the limit, $w(t)$ can be approximated by the Dirac function such that

$$\int_{-\infty}^{\infty} \delta(T - t)f(t)\, dt = f(T) \ .$$

In this case,

$$\int_{-\infty}^{t} \delta(t - T - s)x(s)\, ds = x(t - T) \ .$$

A good example of a delay model for studying chronic leukemia is given by Mackey and Glass (1977).

Another class is a discrete growth model. Let z_t be the population at time t and the population at time $t + T$ be denoted by $z_t + T$. Then we have the difference equation

$$z_{t+1} = z_t F(z_t) = f(z_t) \ ,$$

where $f(z_t)$ is, in general, a non-linear function of z_t. To model a specific type of growth dynamics, we should determine the appropriate form of $f(z_t)$ to reflect known observations or facts about the species in question. One such model which is used frequently is

$$z_{t+1} = z_t \exp\left[a\left(1 - \frac{z_t}{k}\right)\right], \quad a > 0, \ k > 0 \ .$$

Here, $z_t > 0$ for all t if $z_0 > 0$. More information about other types of models of tumor growth are presented in Appendix C.

Chapter 12

Tumor Growth Index

As a rule, tumor growth is described by the number of tumor cells or the volume of the tumor or the mass of the tumor. However, in some cases the patient has several new growths at the same time, which have different morphological structures and sizes, and it may be more convenient to use another characteristic of tumor growth. Here, the tumor growth index (TGI), which is a scalar function of a vector argument, is studied. The TGI allows us to replace the set of characteristics (the number of tumors, the size and morphological structure of each tumor) by a single number. In this chapter we follow the work of Marchuk *et al.* (1989) and Asachenkov (1990). It should be mentioned that this work was done in cooperation with I.N. Martianov, A.N. Kurkin, and B.G. Sobolev (Asachenkov *et al.*, 1987, 1988). Martianov and Kurkin presented clinical data from the Central Clinical Hospital of the Fourth State Office of the Ministry of Public Health of Russia.

12.1 Clinical Data

The disease histories of 1221 patients with intestinal tumors were presented by Martianov and Kurkin from the Department of Proctology of the Central Clinical Hospital of the Fourth State Office of the Ministry of Public Health of Russia.

According to clinical practice, all the new growths detected were removed by surgery (using endoscopy or proctotomy). After surgery, monitoring of the patient and a yearly endoscopy test were organized. It was shown from these data and clinical practices that after the first

Table 12.1. Size and morphological type of new growths for the whole period of study.

Tumor type	Size (cm)						Total
	<0.5	0.5–1.0	1.0–2.0	2.0–3.0	>3.0	Unknown	
Non-epithelial	22	5	5	0	1	0	33
Hyperplastic	570	70	9	0	0	3	652
Adenomatous	1119	420	91	6	2	15	1653
Adenoma with dysplasia of degree I-III	346	262	78	16	0	3	705
Tubular and pileous	50	118	176	46	30	7	427
Cancer I	20	50	65	27	33	4	199
Cancer II	6	2	0	4	15	0	27
Cancer III	1	0	0	1	108	0	110
Cancer IV	0	1	0	0	36	4	41

treatment new tumors with different morphological structures can be detected for some patients. Moreover, the time at which the new growths can be detected is strongly dependent on each individual person. The morphological characteristics of tumors which were removed from this group of patients is presented in *Table 12.1*. In total, 377 malignant and 2470 benign tumors were removed. The majority of the benign tumors are adenomatous new growths (1653). The majority of the malignant tumors are of cancer I type (so-called cancer in situ and focal carcinoma). The cancers which have the biggest size are the tubular and pileous new growth types and cancers II–IV.

There are many reasons for tumor development, but a crucial role is played by local factors: the state of the mucous membrane, the mucous colitis, etc. Polyps develop from hyperplastic ones to adenomatous, tubular, and pileous ones. A polyp can turn malignant at any stage of its development. The transition process from benign to malignant tumor is characterized by the proliferative change in the epithelium, which is called *dysplasia*. After type III dysplasia cancer develops in the polyp. The time taken for the polyp to transform into cancer is about three or four years, or more. The malignant tumors of type I, II, and III differ from each other, in principle, by the size of the tumor and the depth of invasion. Cancer IV is characterized by metastasis to other parts of the body.

Table 12.2. The dependence of the frequency of tumor detection on the type of tumor.[a]

Type	1	2	3	4	5	6	7	8
Frequency	0.07	0.17	0.42	0.17	0.1	0.045	0.02	0.005

[a]In *Tables 12.2* and *12.3* the numbers denoting type represent the following cancers: 1 – non-epithelial new growth; 2 – hyperplastic; 3 – tubular adenoma; 4 – adenoma with dysplasia of I–III degrees epithelium; 5 – tubular and pileous new growth; 6 – cancer I and II; 7 – cancer III; and 8 – cancer IV.

During the testing of the patient the clinician can detect a tumor in any stage of its development. The dependence of the frequency of tumor detection on type and size is presented in *Tables 12.2* and *12.3*.

12.2 Tumor Growth Index

Each tumor can be described by a set of characteristics: morphological type, size, etc. Consider two main characteristics of the tumor – the morphological type and the size. Let $S = \{s_1, \ldots, s_m\}$ be a set of different morphological types of tumor and $R = \{r_1, \ldots, r_k\}$ be a set of tumor sizes in balls (see *Table 12.4*). So, each new growth can be described by two numbers (s_j, r_k), where $s_j \in S$, and $r_k \in R$. Insofar as one patient can have several new growths at the same time, the number of couples of (s_j, r_k) is equal to L.

It is natural to suppose that this set of couples is an estimate of patient state and introduce the tumor growth index as a scalar function of such couples. Unfortunately, it is not possible to apply the approach which was mentioned above (Chapter 6 and Appendix B) to tumor growth when we have several new growths at the same time. This is because it is impossible to identify the state of an organism according to the classification by state traditionally used by clinicians (healthy individuals, a light form of affliction, a serious form, etc.). Here, we consider another approach which does not use an expert's estimate of the state of the patient.

First, we should make one relevant remark. If, for a given patient, we have several new growths with the same morphological characteristics, we will consider such a tumor to be a single tumor with a size equal to the sum of the sizes of the individual new growths. Let n_j be the total number of type j tumors, where $j = 1, \ldots, m$, then let $N = \Sigma n_j$ and p_j, given by $p_j = n_j/N$, be an estimate of the probability of detecting

Table 12.3. The dependence of the frequency of tumor detection on the type and total size of tumors.

Total size	1	2	3	4	5	6	7	8	9	10
Type 1	.4851	.2851	.1447	.0426	.017	.0085	.0043	.0085	.0043	.0
Type 2	.311	.4263	.1528	.0643	.0268	.0134	.0027	.0	.0027	.0
Type 3	.0597	.4242	.1961	.1132	.0606	.041	.0241	.0223	.0205	.0107
Type 4	.1549	.4933	.1644	.0918	.0325	.0287	.0172	.0057	.0076	.0019
Type 5	.077	.2104	.3529	.1086	.095	.043	.0362	.0226	.0271	.0
Type 6	.0633	.2127	.2534	.1267	.2579	.0226	.0181	.0181	.0226	.0
Type 7	.0089	.0	.0	.0089	.973	.0	.0	.0	.0	.0089
Type 8	.0	.0244	.0	.0	.878	.0244	.0	.0	.073	.0

Total size	11	12	13	14	15	16	17	18	≥19
Type 1	.0	.0	.0	.0	.0	.0	.0	.0	.0
Type 2	.0	.0	.0	.0	.0	.0	.0	.0	.0
Type 3	.008	.0053	.0018	.0027	.0018	.0018	.0009	.0027	.0027
Type 4	.0	.0019	.0	.0	.0	.0	.0	.0	.0
Type 5	.009	.0068	.0045	.0023	.0023	.0	.0	.0023	.0
Type 6	.0045	.0	.0	.0	.0	.0	.0	.0	.0
Type 7	.0	.0	.0	.0	.0	.0	.0	.0	.0
Type 8	.0	.0	.0	.0	.0	.0	.0	.0	.0

Table 12.4. Units of size of tumor.

Size of tumor	Value in ball
< 0.5 cm	1
0.5–1.0 cm	2
1.0–2.0 cm	3
2.0–3.0 cm	4
> 3.0 cm	5

the j-type tumor. Let q_{jk} be the number of j-type tumors with size k, and thus $p_{jk} = q_{jk}/n_j$ (where $j = 1, \ldots, m$ and $k = 1, \ldots, K$) is the conditional probability of detecting j-type tumors with size k.

Consider a series of steps or stages through which a tumor may pass sequentially to reach state j: $1 - 2 - \cdots - j - \cdots - m$, with probability $p_{\leq j} = \Sigma p_j$. The probability that the tumor reaches state j and its size is equal to k can be estimated by $p_{\leq j} p_j$. Then, the probability that such a tumor is not detected is $q_j(k) = 1 - p_{\leq j} p_j$. Using *Tables 12.2* and *12.3*, we can calculate $q_j(k)$ (see *Table 12.5*). Now assume that the detection of tumors with different morphological types are independent events. We can write the estimate of the probability of detecting the set of couples in the form

$$p(j, k) = 1 - \prod_{j \in J} q_j(k) \ ,$$

where J is a set of tumor types and $p(j, k) = 0$ if there is no tumor. The function $p(j, k) \equiv \phi \in [0, 1]$ we call the tumor growth index (TGI). The dependence of the values of the TGI on the type and size of tumor is presented in *Figure 12.1*. An example of the calculation of the TGI by the set of couples is given in *Table 12.6*.

12.3 TGI Dynamics

Consider the set of TGI values for the ith patient $\phi(t_1^i), \phi(t_2^i), \ldots, \phi(t_L^i)$, and the function

$$\begin{aligned}
\psi_1^i &= \phi(t_1^i) \\
\psi_{l+1}^i &= \psi(t_l^i) + \phi(t_{l+1}^i) \ , \quad l = 1, 2, \ldots \\
\psi(t) &= \psi(t_l) \quad \text{for} \quad t \in [t_l, t_{l+1}] \ .
\end{aligned}$$

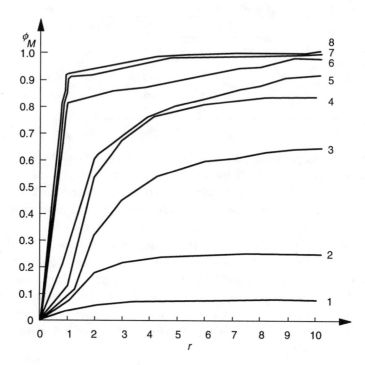

Figure 12.1. The dependence of the function $\phi(j, k)$ on the tumor type j, for $j = 1, 2, \ldots, 8$, and the size.

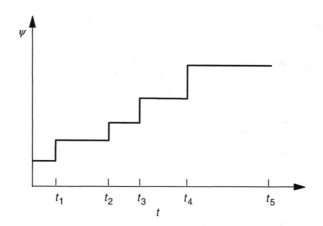

Figure 12.2. An illustration of the intensity of tumor growth as a function of ψ.

Table 12.5. The experimental value of $q_j(k)$.

Total size in ball	Type of tumor[a]							
	1	2	3	4	5	6	7	8
1	.9660	.9254	.8946	.8714	.7424	.1875	.0956	.0790
2	.9461	.8230	.6800	.4620	.3886	.1556	.0827	.0556
3	.9360	.7864	.5512	.3255	.3022	.1328	.0598	.0320
4	.9330	.7709	.4765	.2493	.2328	.1225	.0370	.0170
5	.9318	.7645	.4365	.2224	.1966	.1089	.0168	.0096
6	.9312	.7613	.4094	.1986	.1752	.0868	.0145	.0076
7	.9309	.7606	.3935	.1843	.1415	.0692	.0121	.0056
8	.9303	.7602	.3788	.1795	.1205	.0515	.0097	.0046
9	.9300	.7600	.3653	.1732	.0953	.0295	.0073	.0005
10	.9300	.7600	.3582	.1717	.0886	.0260	.0050	.0000
11	.9300	.7600	.3529	.1708	.0869	.0250	.0050	.0000
12	.9300	.7600	.3494	.1700	.0806	.0250	.0050	.0000
13	.9300	.7600	.3483	.1700	.0764	.0250	.0050	.0000
14	.9300	.7600	.3465	.1700	.0743	.0250	.0050	.0000
15	.9300	.7600	.3453	.1700	.0721	.0250	.0050	.0000
16	.9300	.7600	.3441	.1700	.0719	.0250	.0050	.0000
17	.9300	.7600	.3435	.1700	.0709	.0250	.0050	.0000
18	.9300	.7600	.3417	.1700	.0700	.0250	.0050	.0000
19	.9300	.7600	.3407	.1700	.0700	.0250	.0050	.0000
$\geq$20	.9300	.7600	.3400	.1700	.0700	.0250	.0050	.0000

[a]See footnote to *Table 12.2* for explanation of types.

Table 12.6. An example of a TGI calculation.

Time of inspection	The set of couples	Value of TGI
07.1973	None	0.0000
01.1974	(3,1) (3,1) (3,1)	0.4488
06.1976	(7,5)	0.9832
02.1977	(3,1) (3,1) (3,2) (3,2) (3,2)	0.6212
09.1977	(2,1) (3,2)	0.3708

This function reflects the intensity of tumor growth for a given patient. As a rule, the function has a higher rate of growth for a patient with cancer. An example of the intensity of tumor growth is given in *Figure 12.2*.

According to Burnet's theory, a proportion of normal cells continuously transforms into malignant cells with an individual rate for each patient. As a rule, all tumors detected are removed by surgery. As a result of such an operation the tumor mass is reduced without influencing

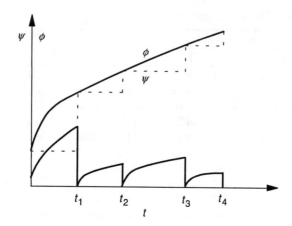

Figure 12.3. An illustration of TGI dynamics.

the mechanism of tumor growth. This means that after surgery it may be months or years before new tumors can be detected.

Suppose that on the time interval $[t_l, t_{l+1}]$, where $l = 1, 2, \ldots$, the TGI dynamics are described by the equation

$$\phi(t) = \phi(t_l) \exp \left(\int_{t_l}^{t} h(s) \, ds \right) \, ,$$

where $h(s)$ is an unknown function. Now, assuming that the tumors detected are not removed, we can write

$$\begin{aligned} \phi(t) &= \phi(t_1) \exp \left(\int_{t_1}^{t_2} h(s) \, ds \right) \times \cdots \times \exp \left(\int_{t_{m-1}}^{t} h(s) \, ds \right) \\ &= \phi(t_1) \exp \left(\sum_{i=2}^{m-1} \int_{t_{i-1}}^{t_i} h(s) \, ds + \int_{t_{m-1}}^{t} h(s) \, ds \right) \end{aligned}$$

or, for all $t > t_1$, $\phi(t)$ will be a monotonous function with

$$\frac{d\phi(t)}{dt} = h(t)\phi(t), \quad \phi(t_1) > 0 \, . \tag{12.1}$$

An example of TGI dynamics is given in *Figure 12.3*.

The analytical form of $h(t)$ should be determined from the available information. To achieve this, display the values of ϕ_l, where $l = 1, 2, \ldots, L$, $L > 3$, for each patient using the following coordinates:

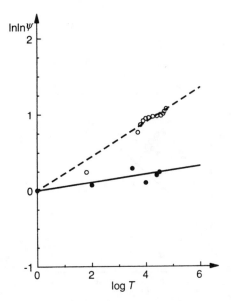

Figure 12.4. Lineal regression. The symbols ● and ○ denote the values of the TGI for different patients.

$(\ln \ln \phi, \ln t)$ (see *Figure 12.4*). We can see that the experimental data are approximated by the following lineal equation:

$$\ln \ln \phi_l = a_0 + p \ln t_l, \quad l = 1, 2, \ldots, L \ . \tag{12.2}$$

This regression was analyzed for 286 patients. The coefficient of correlation, r, was such that $r \geq 0.9$ for 46% of patients, for 40% of individuals $0.7 \leq r \leq 0.9$, and for only 14% r was less then 0.7. So, the function $h(t)$ can be approximated by the Weibull one:

$$h(t) = apt^{p-1}, \quad a > 0, \quad p > 0, \quad \ln a = a_0 \ .$$

It should be mentioned that the parameters of this function can be estimated for each person for whom there are three or more measurements. An example is given in *Figures 12.4* and *12.5*.

12.4 Individual Forecasting of the Time at which New Growths can be Detected

We can use equation (12.1) or (12.2) to forecast the time when the new tumors can be detected. Of course, we should have at least three

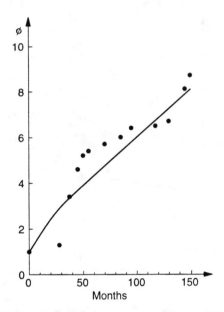

Figure 12.5. Weibull growth: ● are the values of TGI.

measurements of the function $\psi(t)$. Let $y_i = \ln\ln\psi_i$ and $x_i = \ln t_i$, then the mean square estimates are

$$a^* = \overline{y} - p^*\overline{x}$$

$$p^* = \frac{\sum_{i=1}^{L} y_i(x_i - \overline{x})}{\sum_{i=1}^{L}(x_i - \overline{x})^2}$$

$$\overline{y} = \frac{1}{L}\sum_{i=1}^{L} y_i$$

$$\overline{x} = \frac{1}{L}\sum_{i=1}^{L} x_i .$$

Assuming that the measurement errors have a Gaussian distribution with parameters (o, σ^2), the estimate of variance is

$$s^2 = \frac{1}{L-2}\left(\sum_{i=1}^{L}(y_i - \overline{y})^2 - (p^*)^2\sum_{i=1}^{L}(x_i - \overline{x})^2\right) .$$

To estimate the value y_{L+1} at the instant of time t_{L+1} we can use the equation

$$y_{L+1} = a^* + p^* x_{L+1} . \tag{12.3}$$

Table 12.7. Statistical characteristics of the error in the forecast.

E{e}	σ	$\pm m$	min	max	Total number of patients
0.05	0.19	0.01	−0.79	0.86	231

For the $100(1 - \alpha)\%$ confidential interval for mathematical expectation $E\{y_{L+1}\}$ at the instant of time x_{L+1}, we have

$$y_{L+1} \pm t_{L-2}^{0.5\alpha} S(v)^{0.5}$$

where t(.) is a t-statistic and

$$v = \frac{1}{L} + \frac{(x_{L+1} - \bar{x})^2}{\sum_{i=1}^{L} (x_i - \bar{x})^2} \; .$$

The maximum-likelihood estimate for x_{L+1} is

$$
\begin{aligned}
x^* &= \frac{(y^* - \alpha^*)}{p^*} \\
&= \bar{x} + \frac{(y^* - \bar{y})}{p^*} \; .
\end{aligned}
\tag{12.4}
$$

This estimate is biased because in the general case

$$E\{x^*\} \neq \frac{E(y^* - \alpha^*)}{Ep^*} = x^* \; .$$

The appropriate $100(1 - \alpha)\%$ confidence interval has the form $[d_1 + x^-, d_2 + x^-]$, where d_1 and d_2 are solutions of the following equation

$$
d^2 \left((p^*)^2 - \frac{F_{1,L-2}^{\alpha} s^2}{\sum_i (x_i - \bar{x})^2} \right) - 2 d p^2 (y^* - \bar{y})
$$
$$
+ \left[(y^* - \bar{y})^2 - F_{1,L-2}^{\alpha} s^2 \left(1 + \frac{1}{L} \right) \right] = 0
$$

We will have a finite interval if

$$(p^*)^2 > \frac{F_{1,L-2}^{\alpha} s^2}{\sum_i (x_i - \bar{x})^2} \; .$$

The forecast for the individual patient of the time at which the new tumors can be detected, which was given by equation (12.3), has the statistical characteristics which are given in *Table 12.7* and *Figure*

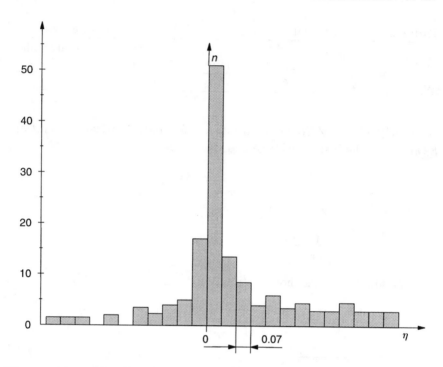

Figure 12.6. The distribution of the error in the forecast.

12.6. Here, $e = \phi_{\text{theor.}} - \phi_{\text{fact}}$ is an absolute error of forecasting. In our experiment, $|e| \leq \sigma$ for 84.4% of patients and $|e| > \sigma$ for only 36 patients out of 231. Analysis of the disease history of patients with the forecast made at the wrong time for detecting new growths shows that for 22 patients we had objective reasons for obtaining the wrong forecast:

- Not all new growths were analyzed.
- Some tumors were lost or completely destroyed during the operation.
- There was no information about the size of tumors.

However, in 14 cases we did not have obvious reasons for forecasting wrongly.

Based on these results, we can say that the TGI can be used for forecasting the time at which new tumors can be detected in individual patients. Unfortunately, we can not apply this method when the number of measurements is less then three. To make a more precise forecast we should use additional information: immunological tests, etc.

Chapter 13

Tumor Dynamics and Control

A foundation for the mathematical modeling of tumors and their control is presented in this chapter. A knowledge-based mathematical model is proposed for the interaction between tumor cells and the immune system. Parametric control variables relevant to the latest experimental data are studied. Sigmoidal dose-response relationships and Michaelis–Menten dynamics are included. The model is composed of 12 first-order, non-linear differential equations based on the cellular kinetics of Chapter 2. Each of the states can be modeled as a bilinear process with non-linear feedback. The preliminary results show that the parametric control variable is important in the destruction of tumors. Also, exacerbation theory is demonstrated for tumor control.

13.1   Introduction

The aim of cancer therapy is to remove or to destroy the tumor without seriously damaging the host. Often, this can be achieved by surgery, radiotherapy, chemotherapy, or immunotherapy, or by a combination of these treatments. Compared to chemotherapy, radiotherapy, and surgery, immunotherapy is the most efficient therapy because immune effector cells naturally kill the target cells without destroying normal neighboring cells.

The potential methods of immunotherapy for tumors can be classified into two broad categories: active, those that attempt to induce in the host a state of immune responsiveness to the tumor; and passive (adoptive), those that directly transfer to the host an immunologically active reagent that mediates an anti-tumor response itself (Lotze and Rosenberg, 1988).

Since the late 1970s, several mathematical models for the interaction between tumor cells and the immune system have been proposed. Rescigno and Delisi (1977) and Grossman and Berke (1980) presented a simple model for the interaction of tumor cells and cytotoxic (killer) T lymphocytes. Lefever and Garay (1978) analyzed the cell-mediated cytotoxic reactions against transformed cells and their negative regulation by blocking factors. Merrill (1983) proposed and analyzed a model of immune surveillance mediated by natural killer (NK) cells. It is well known that the immune response to a tumor involves several effector cells, for example, T lymphocytes, B lymphocytes, macrophages, etc. Most simple kinetic models of the anti-tumor immune response describe only one aspect of the complex phenomena. Therefore, the models proposed above are not sufficient to explain a very complex immune response against a tumor.

Recently, De Boer, Hogeweg, and their associates (De Boer and Hogeweg, 1986; De Boer et al., 1985) presented a model of the macrophage–T lymphocyte interactions that generate an anti-tumor immune response. However, they neglected tumor escape mechanisms and natural killer activity.

Here, we present a mathematical model of the effector mechanism in which the immune system attacks tumor cells through the so-called cell-mediated immune (CMI) response. We follow the work of Mohler and Lee (1990).

13.2   Immunological Foundation

The concept of immunity against established tumors and the related concept of "immunologic surveillance" against emerging, new clones of malignant cells are based on the hypotheses that tumor cells differ antigenically from normal cells, and that the defense mechanisms of the host are capable of recognizing and exploiting these differences.

In order for immunotherapy to be successful, it appears that a tumor must be antigenic. This means that it can stimulate an immune response like other foreign bodies which would be relevant for tumor rejection. If the antigenicity of the tumor is low, immunotherapy will have less effect. A weakly antigenic tumor evokes a weak immune response.

The debris of these tumor cells is phagocytosed by antigen-presenting cells (APCs) that subsequently present antigens in an Ia-restricted fashion to T cells to initiate cell-mediated immunity. Although several types of phagocytic cells may be instrumental in the degradation of antigens, only cells of the mononuclear system can be considered APCs (Barrett, 1988; Erb and Ramila, 1984). All macrophages are not APCs, but only macrophages with the I-A or I-E protein response. The dendritic cell is also capable of antigen presentation.

The stimulation of T lymphocytes by antigen is normally assumed to require at least two signals. Binding of the compound receptor of a T cell to the Ia–antigen complex on the accessory cell constitutes the first signal for the activation of the T cell. To complete this process of activation, the accessory cell must deliver a second signal in the form of lymphokine interleukin 1 (IL-1). Although macrophage and dendritic cells are important sources of IL-1, keratinocytes and other cells may produce IL-1 too (Durum *et al.*, 1985). After being triggered, and concomitant with proliferation, T_h cells release an array of lymphokines with a variety of functions. Some examples are given below.

Macrophage Chemotactic Factor (MCF)

This has a positive, chemotactic effect on monocytes and macrophages causing them to migrate to the site of the T-cell response.

Migration Inhibition Factor (MIF)

The MIF acts on blood-borne monocytes to make them adherent to the endothelial lining of venules. Although the production of MIF is antigen specific, once it is produced the inhibition of a macrophage occurs in a random and non-specific manner, even in the absence of antigen.

Macrophage Activating Factor (MAF)

Lymphokine affects the remaining monocytes, which have migrated to the site of the response and this appears to be an activating factor. The lymphokine's role is the activation of the monocytes into phagocytic cells with increased size, lysosomal granule content, respiratory activity, and ability to ingest particles and debris in the area.

The activation of these cells increases their ability to kill tumor cells. Whole tumor cells are not ingested by macrophages. Macrophages ingest tumor debris following tumor destruction mediated by cytotoxic T cells (T_c), NKs, and activated macrophages, M_a, and antibody-dependent, cell-mediated tumor destruction. Interferons (IFNs) are anti-viral proteins or glycoproteins produced by several different types of cells in the mammalian host in response to viral infection. IFN seems to appear before levels of M_a or Ab are detectable, i.e., it is an early protective device.

It has been suggested that IFN may also play a physiologic role in the regulation of the immune response. It is known that IL-2 induces lymphocytes to produce interferon-γ: this type of IFN is particularly efficient in inducing tumor cell resistance to NK-cell-mediated lysis (Gronberg et al., 1989).

Lymphotoxin (LT)

Class I restricted cytotoxic T cells produce LT in response to non-self class I antigen. It is a unique amino acid sequence that is able to kill certain tumor cells. Interferon may act as an enhancer of LT production in IL-2-activated T cells (where IL-2 denotes interleukin-2). It has some sequence homology with the tumor necrosis factor TNF_β and may have some potential in the treatment of cancer. TNF_α is known to be identical to the hormone cachectin. The activated macrophage and mononuclear phagocytes are the major TNF_α-producing cells (Meager et al., 1989).

Interleukin-2 (IL-2)

The term "interleukin" has been used to describe lymphokine molecules which regulate the activation and proliferation of white blood cells. Major functions of IL-2 are as follows (O'Garra et al., 1988):

- It induces lymphokine production by T cells.
- It induces the growth of activated T cells – thymocytes.
- It induces cytotoxic T-lymphocyte activity.
- It increases natural killer cell activity.
- It increases lymphokine-activated killer cell activity.
- It increases monocyte cytotoxicity.

In addition to the response of activated T cells to IL-2, NK cells also have a major response.

Table 13.1. Humoral and cellular effector immune mechanisms in tu-mor destruction.

Humoral mechanisms	
1	Lysis by Ab and complement
2	Ab- and/or complement-mediated opsonization
3	Ab-mediated loss of tumor cell adhesion
Cellular mechanisms	
1	Destruction by cytotoxic T cells
2	Ab-dependent, cell-mediated cytotoxicity (ADCC)
3	Destruction by activated macrophages
4	Destruction by NK cells

13.2.1   Cellular mechanisms

Humoral and cellular immune effector mechanisms capable of destroying tumor cells are summarized in *Table 13.1* (Benjamini and Leskowitz, 1988). This forms a basis for the following discussion.

Large Granular Lymphocytes (LGLs)

LGLs include NK cells here. Both IL-2 and interferon have been shown to stimulate the growth and activity of LGLs. LGL cells need no previous exposure to the antigens of the target cell before they become active. Target cells are more quickly destroyed by LGLs than by T_c cells and activated macrophages. LGL cells are not antigen specific, although their indiscriminate activity could result from the combined activity of several LGL subsets, each of which has a restricted target-cell specificity. IL-2 alone can promote the growth of human and murine NK cells, in contrast with T cells, in which IL-2 receptors had to be induced by mitogens or antigens for the cells to become responsive to IL-2 (Herberman, 1987).

Antibody-dependent, Cell-mediated Cytotoxicity (ADCC)

ADCC involves the binding of tumor-specific Ab to the tumor surface. The Ab is usually an IgG and may be of any subclass. The effector cell must have surface receptors for the F_c portion of IgG. Note:

- M_ϕ cells and granulocytes (especially neutrophils) have these receptors.

- Platelets and B cells do not participate even though they have F_c receptors.
- Null cells are assumed equvalent to K cells here.

The details and significance of this mechanism for the destruction of tumor cells in vivo is still not completely clear.

Macrophages (M_φ)

Macrophages are indirectly derived from bone marrow promonocytes. After differentiation of the promonocytes into blood monocytes, they settle in the tissues and mature into macrophages. Here, they constitute the reticuloendothelial system. Macrophages may become highly cytotoxic when they become activated by MAFs (Otter, 1986).

Macrophages play an important role in regulating the immune response reaction. They not only act as effector cells against tumors, but also express both positive and negative regulatory effects on humoral and CMI responses during tumor growth. These are the most active of the body's phagocytic cells and are important for antigen processing.

Cytotoxic T cell (T_c)

T cells can be divided into three different subsets: T_c, T_h, and T_s. A viral infection can stimulate a population of killer T cells (T_c) which are specifically cytotoxic for virus-infected host cells that bear viral antigen. Regulation of the CMI response is a complex biological process governed by a series of positive and negative signals. T-helper cells (T_h) are capable of providing the necessary signals which enhance cytotoxic T-cell proliferation. T-suppressor cells (T_s) are characterized by an ability to inhibit the helper function. The network of T_s cells is involved in the regulation of the T_h cell. Suppressor T cells produce soluble factors that mediate suppressive activity. Each T_s subset produces its own type of suppressive factor (Asherson *et al.*, 1986).

The T_c cell recognizes antigen on live, viable, tumor cells in the context of a class I MHC protein. It proliferates in response to T_h-derived IL-2. T_c cells do not require assistance from antigen-processing macrophages of any type. The target cells are destroyed by LTs. LGLs are responsible for "non-specific effects" but T_c cells are thought be instrumental in "specific" resistance (Klein and Vanky, 1981).

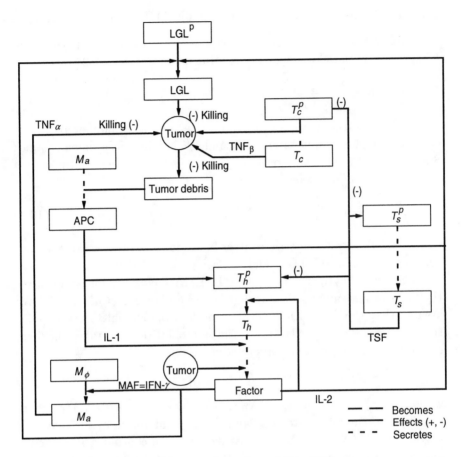

Figure 13.1. Schematic representation of the CMI response.

13.3 The Mathematical Model

From the above discussion we can see that the humoral responses to tumors in vivo are still not well explained. We assume here that humoral mechanisms have no direct role in tumor destruction. Also, the role of the complement is neglected. The block diagram of the whole CMI mechanism is shown in *Figure 13.1*. Based on our knowledge of tumor immunology, we can make a mathematical model of the anti-tumor immune response due to cell kinetics. These cellular kinetics are quite well defined from conservation equations and chemical mass–action principles (see Chapter 2).

13.3.1   Cell-mediated Immune (CMI) Response Model

One widely used, general deterministic tumor growth model studied in Chapter 11 is the Gompertz model (Hanson and Tier, 1982):

$$\frac{dN}{dt} = bN \ln \left(\frac{k}{N} \right) \; ,$$

where $N(t)$ is the measure of tumor size, i.e., the number of tumor cells, k is the maximum tumor size, and $1/b$ is a response time constant. Similarly to the cellular concentration model, a perturbed tumor-cell population is modeled by

$$\frac{dN}{dt} = bN \ln \left(\frac{k}{N} \right) - kill(.)N \; , \tag{13.1}$$

where $kill(.)$, the parametric control function, is a function of the concentrations of T_c, M_a, LGL, and so on.

The typical cytotoxicity against TNF/LT concentration graph takes the form of a sigmoidal dose-response curve (Meager *et al.*, 1989). We assume that the TNF/LT concentration is proportional to the population of corresponding cells. Then, the sigmoidal relation takes the form of the following equation (Carson *et al.*, 1983):

$$CTX = CTX_0 + \alpha_x \tanh \left[\beta_x (x - x_0) \right] \; ,$$

where CTX is cytotoxicity, CTX_0 is given by $(CTX_h + CTX_l)/2$, where CTX_h is maximum cytotoxicity and CTX_l is minimum cytotoxicity, α_x is given by $(CTX_h - CTX_l)/2$, x is the cytotoxin concentration, and α_x and β_x are slopes at x_0, where x_0 is the x value corresponding to CTX_0.

Tumor debris of concentration x_d might be generated to enhance tumor recognition by the immune mechanisms. This would be modeled by the latter term in equation (13.1) with removal time constant τ_d so that

$$\frac{dx_d}{dt} = kill(.)N - \frac{x_d}{\tau^d} \; .$$

The graph of antigen presentation by activated macrophage against antigen concentration follows the Michaelis–Menten limit dynamics (Berzofsky *et al.*, 1988) and might be approximated by

$$z = \frac{z_s x_d}{x_{d0} + x_d} x_a \; ,$$

where x_a is the concentration of M_a, z_s is APC saturation, and x_{d0} is the antigen value corresponding to $z_s/2$.

One of the major activities of IL-1 is to induce the synthesis and secretion of the T-cell-derived, mitogenic lymphokine interleukin-2 (IL-2). This link between IL-1 and IL-2 is an essential element in the T-cell-activation sequence because it involves the conversion of a primary, macrophage-derived maturational signal that results in the amplification of a specific immune response.

As mentioned earlier, IL-2 production by T_h cells requires two signals, i.e., antigen and IL-1. Mizel (1982) shows that the production of IL-2 depends on IL-1: IL-2 is not produced in response to antigen only. The experimental data show that IL-2 is produced when T_h cells are stimulated with IL-1 and mitogen. The synthesis takes the form of a sigmoidal relation. We can approximate the relation between IL-1 and IL-2 by

$$F(.) = \left\{ d_F + (a_F - d_F)/\left[1 + \left(\frac{z}{c_F} \right) k_F \right] \right\} FTH$$

$$FTH = \frac{T_h}{T_{hsat} + T_h} ,$$

where $F(.)$ is the IL-2 concentration, z is the concentration of IL-1, a_F is the minimum IL-2 concentration, d_F is the maximum IL-2 concentration, k_F is a slope parameter, c_F is the concentration of IL-1 which gives 50% of the maximal response, and T_{hsat} is T-helper cell saturation.

Activated T lymphocytes proliferate in response to IL-2 produced by T-helper cells. When an activated T cell divides, it may remain activated or it may return to the resting state. It may depend on the IL-2 receptor expression on the daughter cells. Milanese *et al.* (1987) suggests that the T cell remains in the activated state, and in the absence of additional antigenic stimulation it returns to the resting state. The growth response of T lymphocytes to IL-2 is approximated by a four-parameter logistic function (Hooton *et al.*, 1985):

$$y_T = d_T + \frac{a_T - d_T}{1 + (x/c_T)^{k_T}} ,$$

where y_T is the proliferative response, x is the concentration of IL-2, a_T is the minimum response, d_T is the maximum response, k_T is a slope parameter, c_T is the concentration of IL-2 which gives 50% of the maximal response.

In the healthy state (normal steady state), a constant number of lymphocyte precursor cells are produced each day. However, the influx of precursors is increased due to inflammatory reaction in the cancerous state. Consequently, cell influx is described by:

$$v_i(t) = v_i(1 + p_{0i}) ,$$

where v_i is the cell birth rate in the normal state and p_{0i} is the inflammation rate.

The differentiation of cytotoxic T cell precursors depends on their contact with tumors and IL-2. T-suppressor effector cells may inhibit the generation of precursor cells. The activity of LGL cells can be augmented by lymphokines such as interferon (IFN) and interleukin-2 (IL-2). Macrophages, M_ϕ, may become activated by γ-interferon (a MAF, macrophage activating factor). Several recent studies have suggested that IL-2 regulates the generation of γ-interferon (Dohlsen *et al.*, 1985; Vilcek *et al.*, 1985). The kinetics of γ-INF production by T_h cells stimulated with IL-2 take the same form of sigmoidal relations. Therefore, the relationships between γ-IFN and IL-2 can be approximated by a four-parameter logistic function given by

$$\gamma - INF = d_{INF} + \frac{(a_{INF} - d_{INF})}{1 + (F/c_{INF})^k} \ INF .$$

The differential equations of the model which are considered above are given in the following form:

$$\frac{dx_1}{dt} = v_1(t) - \frac{x_1}{\tau_1} - p_{16}x_1 - p_{81}x_1$$

$$\frac{dx_2}{dt} = v_2(t) - \frac{x_2}{\tau_2} - p_{27}x_2 - p_{27}x_2x_{11} - p_{82}x_2$$

$$\frac{dx_3}{dt} = v_3(t) - \frac{x_3}{\tau_3} - p_{38}x_3 - p_{83}x_3$$

$$\frac{dx_4}{dt} = v_4(t) - \frac{x_4}{\tau_4} - p_{49}x_4 - p_{49}x_4$$

$$\frac{dx_5}{dt} = v_5(t) - \frac{x_5}{\tau_5} - p_{5,10}x_5$$

$$\frac{dx_6}{dt} = p_{16}x_1 - \frac{x_6}{\tau_6} + p_6x_6$$

$$\frac{dx_7}{dt} = p_{27}x_2 + p_{27}x_2x_{11} - \frac{x_7}{\tau_7} + p_7x_7$$

$$\frac{dx_8}{dt} = p_{38}x_3 - \frac{x_8}{\tau_8} + p_8x_8$$

$$\frac{dx_9}{dt} = p_{49}x_4 + p_{49'}x_4 - \frac{x_9}{\tau_9}$$

$$\frac{dx_{10}}{dt} = p_{5,10}x_5 - \frac{x_{10}}{\tau_{10}}$$

$$\frac{dx_{11}}{dt} = p_{11}x_{11} - \frac{x_{11}}{\tau_{11}} - p_{7,11}x_{11} - p_{9,11}x_{11} - p_{10,11}x_{11}$$

$$\frac{dx_{12}}{dt} = \frac{x_{11}}{\tau_{11}} - \frac{x_{12}}{\tau_{12}} + p_{7,11}x_{11} + p_{9,11}x_{11} + p_{10,11}x_{11} \ .$$

Here, x_i, where $i = 1, \ldots, 12$, is the state of each cell at a certain time instant. The explanation of the terms in these equations is as given below.

- The subscripts are as follows.

  $1 - T_h^p$, $2 - T_c^p$, $3 - T_s^p$,

  $4 - LGL^p$ (where p indicates "precursor")

  $5 - M_0$, $6 - T_h$, $7 - T_c$, $8 - T_s$

  $9 - LGL$, $10 - M_a$, $11 - $ tumor, $12 - $ tumor debris .

- The definitions of other terms are as follows.

  Cell influx: $v_i(t) = v_i(1 + P_{0i})$, $i = 1, \ldots, 5$

  Inflammation rate: $p_{0i} = p_{id} + (p_{ia} - p_{id})/\left[1 + (F/c_i)_i^k\right]$

  IL-2: $F(.) = [d_F + (a_F - d_F)/(1 + z/c_F)][x_6/(T_{hsat} + x_6)]$

  IL-1: $z = F_D(.)x_{10}$, $F_D(.) = (z_s x_{12})/(x_{d0} + x_{12})$

  $\gamma - INF : \gamma - INF = d_{INF} + (a_{INF} - d_{INF})/\left[1 + (F/c_{INF})^k INF\right]$

  $p_{16} = \mu_1(1 + z)$, $p_{27} = \mu_2$, $p_{27'} = \mu_2 F(.)$, $p_{38} = \mu_3(1 + F(.))$

  $p_{49} = \mu_4(1 + F(.))$, $p_{49'} = \mu_4 * \gamma - INF$, $p_{5,10} = \mu_5 * \gamma - INF$

  $p_{81} = p_{82} = p_{83} = T_s F$

  $$= d_{TSF} + (a_{TSF} - d_{TSF})/\left[1 + (x_8/c_{TSF})^k TSF\right]$$

  $p_6 = p_7 = p_8 = d_T + (a_T - d_T)/\left[1 + (x/c_T)\right]^k T$, $p_{11} = b \ln(k/x_{11})$

  $p_{7,11} = p_{9,11} = p_{10,11}$

  $$= CTX, \quad CTX = CTX_0 + \alpha_x \tanh[\beta_x(x - x_0)] \ .$$

- Assume that the inflammation rates for all immune effector cells are identical.
- Initial conditions:

$$
\begin{array}{llll}
x_1(0) &=& 1.05 \times 10^8 & \qquad x_7(0) &=& 1.1 \times 10^6 \\
x_2(0) &=& 2.2 \times 10^7 & \qquad x_8(0) &=& 1.1 \times 10^6 \\
x_3(0) &=& 2.2 \times 10^7 & \qquad x_9(0) &=& 1.56 \times 10^6 \\
x_4(0) &=& 1.3 \times 10^7 & \qquad x_{10}(0) &=& 6.5 \times 10^4 \\
x_5(0) &=& 1.3 \times 10^7 & \qquad x_{11}(0) &=& 1.0 \\
x_6(0) &=& 5.25 \times 10^6 & \qquad x_{12}(0) &=& 0.001 \ .
\end{array}
$$

Parameter values

Ideally, experimental data should be obtained under consistent conditions which match the problem simulation. Unfortunately, it is very difficult to obtain consistent data. Here, the parameter values are obtained from many sources using different experiments which are often not very consistent. Several parameters are, however, still unknown. Therefore, these parameter values were chosen somewhat arbitrarily.

A typical thymus contains about 200 million cells. The thymic cortex produces about 50 million cells each day, most of which will disappear within 3 days, i.e., $\tau_i \approx 3$ days, $i = 1, 2, 3$. It is assumed that 30% of these cells are T_s and 70% are T_h (Scollay and Shortman, 1985). Activated T lymphocytes are considered to be cells with a long lifetime. The turnover time of T cells is assumed to be 50 days (Grossman and Berke, 1980).

It has been shown that bone marrow is required for the proliferation and differentiation of natural killer (NK) cells and that these cells have a life-span of a few weeks. In human peripheral blood, NK phenotype cells represent an average of 15% of the cells with lymphoid characteristics, with large individual variations (Trinchieri *et al.*, 1987). The total number of T cells in the blood is of the order of 10^8 cells. The influx of NK cells is, therefore, about 750,000 cells per day.

Blood monocytes originate in the bone marrow from dividing precursor cells. They then enter the peripheral blood, in which they circulate until they leave it to become macrophages in cell tissue. Normal macrophages in cell tissue consist of about 1.5×10^7 cells (Van Furth *et al.*, 1982). This is almost consistent with the macrophage concentration of one per 10 or 100 T cells in vitro (Kevrekidis *et al.*, 1988). The calculated mean turnover time of a macrophage is about 20 days. The influx of macrophages is, therefore, about 750,000 cells per day. The turnover time of activated macrophages is assumed to be short, about one day (De Boer *et al.*, 1985).

During the inflammatory reaction, the influx of immune effector cells is assumed to increase 10 times. The parameter values for the simulation are shown in *Table 13.2*.

13.4   Simulation and Analysis

Experimentation shows that the multiplicative (or parametric) control of the tumor by means of individual immune mechanisms results in the death rate of the tumor being increased by immune effector cells. The slope of the tumor death rate curve is so steep that we cannot solve effectively a differential equation of this immune model using conventional numerical integration methods (e.g., explicit Runge–Kutta methods). The system becomes a "stiff" differential equation. As a means of solving stiff systems in general, the most commonly used methods are semi-implicit Runge–Kutta and Gear methods. Here, the IMSL routine DGEAR is used to integrate the ODEs.

First, the effects of the antigenicity of tumors were investigated. As mentioned earlier, higher antigenic tumors stimulate the T_h cells, which produce enough IL-2 to make it necessary to use effector cells to remove target cells. *Figure 13.2(a)* shows that a tumor which has one cell at the initial instant of time grows exponentially until sufficient effector cells appear and then it regresses rapidly. Then there is no reason for the number of immune effector cells to remain so high. Therefore, the number of effector cells decreases according to the decrease in the size of tumors. In the case of lower antigenicity, the amount of IL-2 is not sufficient to stimulate the immune system so the concentrations of effector cells are essentially the same as in the healthy state. A tumor can no longer be rejected and there is a breakthrough phenomenon as shown in *Figure 13.2(b)*. IFN is also very important in removing tumor cells. *Figure 13.2(c)* shows that the tumor grows progressively when no IFN is produced. The population of M_a decreases exponentially in this case.

In tumor immunology, the final goal is tumor control. Among several methods of immunotherapy, exacerbation theory can be useful for tumor treatment. *Figure 13.2(d)* shows the tumor regression due to an initial increase of effector cells. During the decrease in effector cells, the tumor re-occurred and reached the equilibrium state. At 66 days, 10^3 tumor cells are injected over a period of 2 weeks. This higher tumor

Table 13.2. Parameter values of the model.

Parameter	Units	Explanation
$\tau_i = 3 \; (i = 1,2,3)$	days	Lymphocyte death time
$\tau_4 = 20$	days	LGL^p death time
$\tau_5 = 20$	days	Macrophage death time
$\tau_i = 50 \; (i = 6,7,8)$	days	T cell death time
$\tau_9 = 20$	days	LGL death time
$\tau_{10} = 1$	days	Angry macrophage death time
$\tau_{11} = 1000$	days	Tumor death time
$\tau_{12} = 0.5$	days	Tumor debris removal time
$\mu_i = 0.001 \; (i = 1, \ldots, 5)$	cells/day	Activation rate
$v_1 = 3.5 \times 10^7$	cells/day	Lymphocyte birth rate
$v_2 = 7.5 \times 10^6$	cells/day	Lymphocyte birth rate
$v_3 = 7.5 \times 10^6$	cells/day	Lymphocyte birth rate
$v_4 = 750,000$	cells/day	LGL birth rate
$v_5 = 750,000$	cells/day	Macrophage birth rate
$1/b = 9.35$		e-folding time
$k = 2.93 \times 10^{10}$	cells	Maximum tumor size
$CTX_h = 10$		Maximum cytotoxicity
$CTX_l = 0$		Minimum cytotoxicity
$\beta_x = 0.2$		Slope at x_0
$z_s = 1$		APC saturation constant
$x_{70} = 7 \times 10^8$		T_c corresponding to CTX_0
$x_{90} = 3 \times 10^7$		LGL corresponding to CTX_0
$x_{100} = 1 \times 10^6$		M_a corresponding to CTX_0
$x_{d0} = 5 \times 10^5$		Antigen value corresponding to $z_s/2$
$d_F = 1,000$		Maximum IL-2 concentration
$a_F = 0.01$		Minimum IL-2 concentration
$k_F = 1$		Slope parameter
c_F		Varies according to antigenicity
$d_{INF} = 40$		Maximum INF concentration
$a_{INF} = 5$		Minimum INF concentration
$c_{INF} = 10$		Concentration of IL-2 which gives 50% of the maximal response
$k_{INF} = 1$		Slope parameter
$p_{id} = 10 \; (i = 1, \ldots, 5,13)$		Maximum response
$p_{ia} = 0 \; (i = 1, \ldots, 5,13)$		Minimum response
$k_i = 1 \; (i = 1, \ldots, 5,13)$		Slope factor
$c_i = 10$		Concentration of IL-2 which gives 50% of the maximal response
$a_T = 0$		Minimum response
$d_T = 1$		Maximum response
$k_T = 1$		Slope parameter
$c_T = 10$		Concentration of IL-2 which gives 50% of the maximal response
$d_{TSF} = 9$		Minimum response
$a_{TSF} = 0$		Maximum response
$k_{TSF} = 1$		Slope parameter
$c_{TSF} = 107$		Concentration of t_s which gives 50% of the maximal response
$T_{hsat} = 5 \times 10^8$		T_h cell saturation

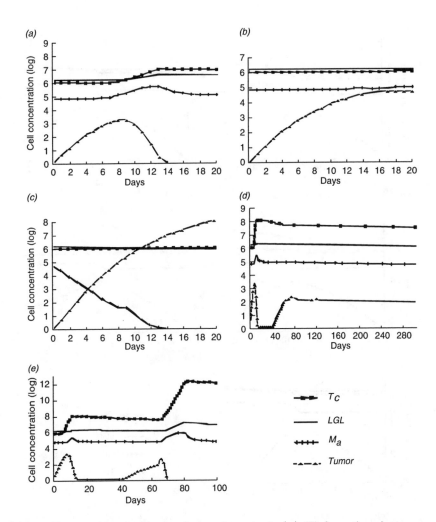

Figure 13.2. Tumor dynamics and control. (*a*) High antigenic tumor. (*b*) Low antigenic tumor. (*c*) No interferon. (*d*) Tumor dynamics. (*e*) Exacerbation theory.

concentration evokes the stimulation of effector cells and the tumor is destroyed completely [*Figure 13.2(e)*].

Recent reports have demonstrated that anti-tumor activity was observed on treatment with either a high dose of IL-2 alone or a lower dose of IL-2 (due to the toxicity of the IL-2) in combination with lymphokine (IL-2)–activated killer (LAK) cells: see Cameron *et al.* (1988)

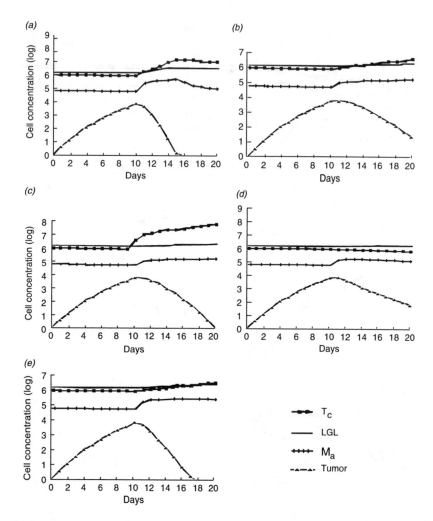

Figure 13.3. Adoptive immunotherapy. (*a*) High IL-2 dose. (*b*) Low IL-2 dose. (*c*) Low IL-2 and T_c dose. (*d*) Low interferon dose after 10 days. (*e*) Low IL-2 and *INF* dose at 10 days.

and Rosenberg *et al.* (1988). They have also shown that the administration of IL-2 and IFN produces a substantial, synergistic, therapeutic effect and that this effect is synergistic with that of tumor-infiltrating lymphocytes (TIL).

Figure 13.3(a) shows that the administration of high doses of IL-2 against a weakly antigenic tumor after 10–15 days can mediate tumor regression. The anti-tumor activity of a combination of therapies using low doses of IL-2 and anti-tumor effector cells such as lymphokine–activated killer cells or tumor-infiltrating lymphocytes is simulated against a weakly antigenic tumor [*Figures 13.3(b, d, e)*]. *Figure 13.3(c)* demonstrates the synergistic, anti-tumor effects of combination therapy with low doses of IL-2 and T_c cells after 10 days [compare with *Figure 13.3(b)*].

Chapter 14

Data Analysis in Cancer Research

The state of an organism during disease is assessed according to clinically measured laboratory indices. The disorders in the normal functioning of the main homeostatic systems which are caused by the disease lead to deviations of these indices from the values which correspond to a healthy body state. Checking the state of the organism during treatment based on the objective analysis of available information is an indispensable pre-requisite for the right choice of treatment tactics.

14.1 Motivation

The main problem in using laboratory data for the assessment of the state of a patient lies in the interpretation of the individual values of indices which characterize the functional activity of physiological systems. In 1975, some special functions, called graveness or generalized indices (GI), were proposed by G.I. Marchuk for analyzing the course of such acute, infectious diseases as viral hepatitis, pneumonia, etc. (see Part II, Chapter 6). Experience in utilizing these functions has shown that they can be successfully used in solving such practical problems as assessing the patient's state, analyzing the process dynamics, and comparing different treatment methods as far as their effectiveness is concerned. However, use of the GI for the assessment of the state of oncological patients has proved impossible. Looked at from one respect, it is not possible to isolate an acute phase of disease when the parameters characterizing that process essentially deviate from their homeostatic values.

247

However, in a different respect, it is difficult to isolate patients with a
disease having a "smooth" course without aggravations and ending in
a complete clinical recovery. This outcome – complete clinical recovery
– is a typical picture of most infectious diseases. All this calls for the
development of new mathematical methods of clinical data processing
which take into account the specific features of an oncological process.

Let us discuss some aspects of the oncological disease that are important for mathematical modeling and for the development of methods
of clinical data processing.

- Tumor growth and metastatic spreading develop without any pronounced clinical effects during a prolonged period of time (months
  or years). One can judge the character of the progress of the disease using only clinically measured laboratory indices which, as is
  known, are only a reflection of local tissue processes. The main
  result of the growth of the tumor is functional disturbances of the
  main physiological systems of the organism. In turn, the organism,
  responds to the development of the neoplastic process by means of
  physiological and compensating reactions. These reactions provide
  stability in the basic physiological functions of the organism and
  guarantee neutralization of infrequent and random disturbances of
  the homeostatic system. Systematic disturbances in most systems of
  the organism for long periods of time lead to considerable structural
  and functional disturbances (Dilman, 1983).
- To characterize the disease and to assess the efficiency of treatment
  in oncology, criteria related to the duration of the process are utilized: for example, the duration of the period without relapse and
  the duration of the patient's life after radical surgery.
- The development of the oncological process brings about systematic
  disorders in the functioning of most systems in the organism, which
  lead to major structural and functional shifts in homeostasis, raising
  the probability of the patient's death. Among the main causes of
  death the following can be distinguished:

  – Mechanical tissue damage of those organs in which the tumor
    metastases result in the operational deficiency of the parenchy-
    mal organs (heart, liver, and lungs) caused by resorption into
    the blood of toxic agents secreted by necrotic elements of the
    tumorous tissue.

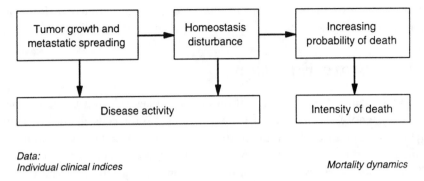

Figure 14.1. Main characteristics of disease.

    — The change in metabolic processes (shifts in the water–salt metabolism, in the acid–base ratio, and metabolic immunode-pression) due to diffuse intoxication leads to the dystrophy of organs and consequent disorders in their functioning. The cascade of such disorders augments secondary intoxication. Tumorous tissue also discharges factors that augment the thromboembolic risk, which in turn induces metabolic shifts.

Thus, the progress of a disease is a pathophysiological process that causes a systematic disorder of the homeostasis which acquires an unstable and uncompensated character that considerably raises the probability of the patient's death. If we consider the dynamics of the tumor disease for a group of patients as trajectories in the space of the parameters, we can see breaking trajectories (i.e., with sudden changes). The instants at which the trajectories break show some stable behavior depending on age, sex, stage of illness, method of treatment, etc.

This general definition permits us to construct a model of the course of the oncological disease as it links two of its most important properties: the rate of growth of its functional disorders (i.e., the activity of the disease) and the intensity of the decrease in the number of patients surviving after radical treatment over a certain period of time (*Figure 14.1*).

Indeed, when assessing the activity of the disease using a set of clinically checked indices, the sum of the individual trajectories of physiological indices corresponds to the survival function. Thus, the probability of death can be determined by the statistical characteristics of the distribution of those trajectories. Note that only the group of mortality characteristics can be accurately observed, whereas homeostatic disorders can be assessed only indirectly using the available clinical data.

14.2   State Equation

Let us consider the dynamics of observed indices relating to cancer patients. Denote the instant of time at which tumor cells begin to occur by t_H and denote the number of cancer cells by $z(t)$. According to clinical practice, the detected tumor is removed by surgery at time t_0. Since radical surgery leads to a decrease only in the mass of the tumor and does not generally affect the cancer process, some remaining tumor tissue continues to grow. As is known, in response to the growth of the tumor, the immune system reacts by producing specific cells and molecules. These components of the immune system can distinguish and destroy cancer cells, at least under ideal conditions. Denote $x(t) \in \mathbb{R}^n$ as a measured vector composed of clinical indices. The development of cancer leads to deviations in these indices from the values corresponding to the healthy organism (no tumor). After "successful" surgery, the indices tend to values which correspond to a healthy state.

The disease activity is determined by a balance between the influence of the tumor on the systems of the organism and the anti-tumor resistance of the organism, that in its turn determines the time at which the tumor reaches a critical state in its development which will lead to the inevitable death of the organism. The disease activity, measured as the intensity of homeostatic disorder, can be assessed by the pattern of variations in the clinically measured indices.

The questions which are studied in this section deal with the equations which describe the dynamics of the measurable indices. The crucial step in this direction is to single out the group of patients with a "favorable clinical history" and to construct a so-called "reference trajectory".

14.2.1   Average dynamics and reference trajectory

It is difficult and sometimes impossible to single out a group of cancer patients with a "smooth" disease history that results in the restoration of the affected system with complete clinical recovery. However, it is possible to single out a group of patients with a "favorable clinical history". This means that the maximum amount of time of the patient's life after the beginning of treatment may be observed.

Let the dynamics of the clinically measured indices for the patients with a life-span of five years or more be described by the equation

$$\frac{dx(t)}{dt} = f(x, \alpha), \quad x(0) = x_0 \geq 0, \quad t \geq 0$$

$$x \in \mathbb{R}^n, \quad \alpha \in \mathbb{R}^L . \tag{14.1}$$

If the measurements are carried out with a group of patients which has A_0 members, we have a set of values

$$\chi_0 = \{x(t_k) : t_k \in [0, T], \ k = 1, \ldots, m_i ,$$
$$i = 1, \ldots, A_0, \ x \in \mathbb{R}^n\} ,$$

where the t_k values, $k = 1, \ldots, m_i$, are the instants at which measurements were taken and A_0 is the number of patients in the group.

The problem under consideration is to find the analytical form of the function $f(x; \alpha) : \mathbb{R}^n \times \mathbb{R}^l \to \mathbb{R}^n$ by χ_0. If we do not have any additional information about the process in question (compare with the model from Chapter 13), it seems reasonable to choose the function $f(x; \alpha)$ so that it has the polynomial form

$$f^j(x; \alpha) = \alpha_0^j + \sum_{i=1}^{n} \alpha_i^j x_i + \sum_{i,k=1}^{n} \alpha_{ik}^j x_i x_k + \cdots$$

$$j = 1, \ldots, n .$$

Now the methods from Part II can be used to estimate the coefficients and degree of the polynomial.

Definition
A solution of the equation (14.1) – $x(t, \alpha^*)$ which describes the average trajectory in the group of patients with favorable clinical histories – is called a support solution or a reference trajectory, and the vector α^* is called a reference or support vector.

Example 14.1
For patients with stomach cancer (Marchuk *et al.*, 1989; Asachenkov *et al.*, 1989) the following patterns have been used. (Clinical data were presented by N.N. Vasilyev and Ye.S. Smolianinov from the Research Institute of Oncology in the city of Tomsk, Russia.) For the percentage of lymphocyte, the equations are

$$\frac{dx(t)}{dt} = 0, \quad x(0) = \alpha^* > 0 .$$

Table 14.1. Coefficients and initial values for reference trajectories.

Variable	Coefficients		Initial conditions
x_1	$p_0=0.3983$	$p_1=0.1343$	6.29
x_2	$p_0=0.2464$	$p_2=0.0062$	2.5
x_3	$p_0=0.2583$	$p_1=0.1505$	4.13
x_4	$p_0=0.1042$	$p_2=0.0277$	0.53

For the concentration of B cells and IgG, the equations are

$$\frac{dx(t)}{dt} = \alpha_0{}^* - \alpha_1{}^*x(t), \quad x(0) = x_0 \geq 0$$

and for the quotients T cell/B cell and IgA/IgM,

$$\frac{dx(t)}{dt} = \alpha_0{}^* + \alpha_1{}^*x(t) - \alpha_2{}^*x^2(t), \quad x(0) = x_0 \geq 0 \ .$$

The reference trajectories are presented in *Figure 14.2*. The coefficients and initial values $x_i(0)$, where $i = 1,\ldots,4$, which have been estimated by using the algorithms and programs developed in the Institute of Numerical Mathematics of the Russian Academy of Sciences, are presented in *Table 14.1* (see Part II, Chapter 8). Here, x_1 is the concentration of B cells, x_2 is the quotient T cell/B cell, x_3 is the concentration of IgG, and x_4 is the quotient of IgA/IgM ($p_1 \ll p_2$ for x_2 and x_4).

Our experience leads us to conclude that after "successful" surgery in the group of patients with favorable clinical histories we can see the "recovery process" when the indices x_i, where $i = 1,\ldots,n$, tend to their normal homeostatic values, $x_i \to x_\infty$ as $t \to \infty$.

14.2.2   Stochastic model

Consider the individual trajectories for the patients with favorable clinical histories. Let $X_m = \{x^j(t), t \in \theta^j, j = 1,\ldots,m\}$, that is, a trajectory's set of indices measured at the instants of time θ_j given by $\theta^j = \{t_1^j,\ldots,t_{nj}^j\}$, where j has values $1,\ldots,m$ and is the number of patients in the group. If the values $x_j(t)$ at $t \in \theta^j$, where $j = 1,\ldots,m$, are connected by straight lines, we have what we call a trajectory bunch. A typical situation is given in *Figure 14.3*. The real trajectories presumably have a stochastic character and cannot be modeled conveniently within the framework of a deterministic model. However, their average dynamics is described by a deterministic equation. We can say that all

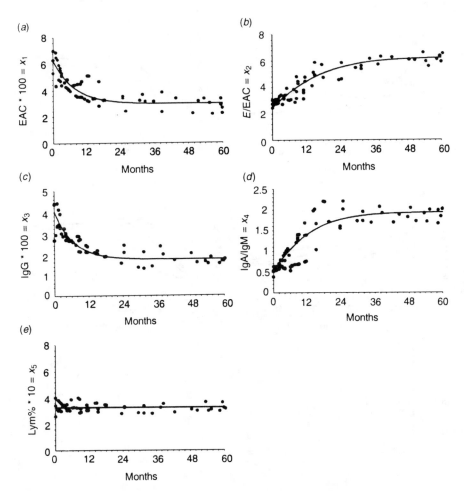

Figure 14.2. Reference trajectories for the patients with favorable disease histories: (*a*) EAC-POK (B cells); (*b*) relation E-POK/EAC-POK (T cells/B cells); (*c*) IgG; (*d*) relation IgA/IgM; (*e*) percentage of lymphocytes. Dots indicate the actual data.

these trajectories have a common law of behavior and the differences between them have a random character.

14.2.3   The sources of perturbations

The main source of perturbations involves small, random deviations of the alpha coefficients from α_0 due to the individual singularity of the organism, i.e., $\delta\alpha = \alpha - \alpha_0$. Alternatively, the source of the perturbations

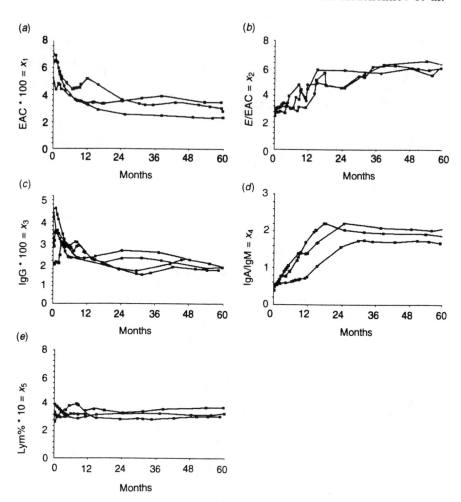

Figure 14.3. Examples of individual trajectories for patients with favorable clinical histories: (*a*) concentration of B cells; (*b*) ratio of percentage of T cells to percentage of B cells; (*c*) IgG; (*d*) IgA/IgM; (*e*) percentage of lymphocytes. The term t is time in months.

may be continuous perturbations of the model coefficients by stochastic processes which reflect the irregular diffusion influence of different factors on the organism. Also, the source of the perturbations may be a random deviation of the state variables at the instant of time $t = 0$: for example, $x_0 \sim N(m_0, \gamma_0)$, i.e., a normal distribution with mean m_0 and variance γ_0.

As a rule, the fluctuations of the coefficients of the model are due to the influence on the process of numerous factors. The central limit theorem states, roughly, that for any distribution pattern of the factors which determine the development of the process, if the effect of their influence is not powerful, and if these factors are independent, the perturbations have a Gaussian pattern in the limit. Hence, to construct a convenient perturbation model, we choose the model of fluctuation within the most-studied class of random processes, i.e., the Gaussian one. Note, however, that the lognormal or Poisson distribution does offer some physical basis for the process at hand.

14.2.4 Model

As a mathematical model for computer simulation, an ODE with random parameter perturbations might be used (Zuev, 1988). (For further details see Part II, Chapters 6 and 7.) Thus,

$$\frac{d}{dt} x_t^\varepsilon = f(x_t^\varepsilon, \alpha^* + \xi_{t/\varepsilon})$$
$$x_0^\varepsilon = x_0, \ t \in [0, T] \ .$$

Here, x_t^ε is a perturbed solution, ε is a low-valued parameter where $\varepsilon > 0$, ξ_t is a stochastic process such that $E\xi_t = 0$ and $\mathrm{cov}(\xi_t, \xi_{t+\tau}) \to 0$ as $\tau \to \infty$.

Again, the trajectory might be considered to be the result of a small perturbation of a dynamic system. The perturbed motion is described by this model as the fast, random fluctuation along the reference trajectory $x_t(\alpha^*)$. In this connection, m trajectories obtained from a homogeneous group of patients can be regarded as m solutions of a stochastic differential equation (SDE). Let $Y_t^\varepsilon = x_t^\varepsilon - x_t(\alpha^*)$ be a deviation between the perturbed motion and the reference trajectory. Then, as is given by Zuev, under certain assumptions with respect to the right-hand side and ξ_t (Ventsel and Freidlin, 1979), the process Y_t^ε is approximated by the linear stochastic differential equation

$$dY_t = \frac{\partial}{\partial x} f(x_t, \alpha^*) Y_t \, dt + \frac{\partial}{\partial \alpha} f(x_t, \alpha^*) \, dw_t$$
$$Y(0) = 0, \ t \in [0, T] \ , \tag{14.2}$$

where x_t is a support solution and w_t is a Gaussian process with independent increments such that $Ew_t = 0$ and $\mathrm{cov}(w_t, w_t) = \Gamma t$, where Γ is the matrix of intensities of ξ_t.

This is the basic equation for our research. Equation (14.2) reflects our knowledge about the process in question. The functions $a(t) = \partial f(x, \alpha)/\partial x$ and $b(t) = \partial f(x, \alpha)/\partial \alpha$ depend on the right-hand side of the model. Rewrite equation (14.2) in a more convinient form:

$$dY_t = a(t)Y_t dt + D(t)dw_t \qquad (14.3)$$

where w_t is a Wiener process. Therefore the matrix $D(t)$ satisfies the condition $DD^T = b\Gamma b^T$.

14.3 Mortality Dynamics

In the previous section the dynamics of observed indices in the group of patients with favorable disease histories are given. What differentiates the dynamics of these indicators in the groups of patients with different life-spans after surgery? How can we estimate the activity of the remaining tumor mass and inspect it during treatment for individual patients? At the present time it is not possible to predict the individual sensitivity of patients to methods of treatment or to inspect the tumor growth process during treatment using conventional statistical methods for processing the clinical data. This is one of the actual problems of clinical oncology.

Consider a group of individuals after radical surgery. The physiological status of these patients can be represented by a point in m-dimensional state space. This state space is specified by the set of measurable indices given by

$$X_t = (x_t^1, \ldots, x_t^n)^T, \quad n \geq 1 \ .$$

An individual response generates the trajectory in this state space:

$$x(t) \equiv x_t = (x_v \in \mathbb{R}^n, 0 \leq v \leq t) \ .$$

Assume, for simplicity, that we deal with just one cohort (a closed group of patients all of which were studied from the same instant of time) and there exists a so-called failure or termination time T for each individual trajectory. This failure is associated with death or specific health changes which enable the failure time to be determined precisely. We have the following requirements:

- A time origin must be unambiguously defined.
- A scale for measuring the passage of time must be agreed.

- The meaning of failure must be entirely clear (Cox and Oakes, 1984).

The time origin need not be, and usually is not, the same calendar time for each individual. Each patient's failure time is measured from their date of surgery or the beginning of treatment. Further, assume that the time-span of the process in question is maximal for all trajectories: the termination time T_{max} satisfies the inequality $T_{max} < \infty$. This means that we shall consider the process on the interval $[0, T_{max}]$.

If we consider the trajectories of $x(t)$ for the patients with different life-spans after radical treatment, we can see the breaking trajectories. The instants at which a break occurs demonstrate stable behavior depending on age, sex, method of therapy, and so on. The health changes in certain cohorts can be represented by a set of individual trajectories given by

$$(x_t(w_t), i = 1, \ldots, N)$$

and a corresponding set of termination times given by

$$\Theta = \{T_1, \ldots, T_N\} \ ,$$

where N is the sample size. In principle, we can derive the frequency distribution at each instant of time t from these N individual trajectories as $N \to \infty$. Hence the trajectory set generates the probability density function of failure times, and we can study the relation between the trajectory set and the probability distribution of the random failure time.

Let T be a non-negative, continuous, random variable (termination time) with a probability density $f(t)$. Then the distribution function $F(t)$ defines the probability that an individual trajectory will fail at or before time t. It is possible to define a continuous function of time,

$$s(t) = 1 - F(T) \ ,$$

which represents the probability that the individual will survive until time t. The function $s(t)$ is called a survival function. If the probability density $p(t)$ exists, then

$$s(t) = \int_t^\infty p(v) \, dv, \quad s(0) = 1 \ .$$

It should be noted that $s(t_1) \geq s(t_2)$, $t_2 \geq t_1$, is a monotonic, decreasing function of time.

Another important characteristic in survival analysis is the so-called mortality intensity $\lambda(t)$, which is determined by the expression

$$\lambda(t) \;=\; \lim_{\Delta t \to 0} \frac{1}{\Delta t}\, Pr\,\{\text{failure in interval } (t, t+\Delta t)$$
$$\text{given survival until time } t\} \;\;.$$

It is simple to obtain the form of the failure rate equation for intensity as

$$\lambda(t) = -\frac{1}{s(t)}\frac{d}{dt}\, s(t) = -\frac{d}{dt}\,\ln\, s(t) \;\;.$$

Then the mortality dynamics in a group are described by the survival function

$$\frac{d}{dt}\, s(t) \;=\; -\lambda(t)s(t), \quad s(0) = 1$$
$$s(t) \;=\; \exp\left(-\int_0^t \lambda(u)du\right), \quad S(0) = 1 \;\;.$$

The total mortality intensity in the interval $[0, T]$ is

$$\Lambda(t) = \int_0^t \lambda(s)\, ds = -\ln\, s(t), \quad \Lambda(0) = 0 \;\;.$$

This may be interpreted as a pathological pressure upon the organism caused by the disease up to the instant of time t.

So, the rate of change in the survival function at t is represented as the product of two independent factors: the failure rate and the survival function at t. We can write the probability density function in the form

$$p(t) = \lambda(t)s(t) \;\;.$$

These results show that specifying the failure rate is sufficient to find the functional form of the survival and density functions. There are many reasons why consideration of the failure rate function may be a good idea, for example:

- It may be intuitively enlightening to consider the immediate risk attached to an individual known to be alive at time t.
- Comparisons of groups of individuals are sometimes more decisively made via the failure rate function.
- Failure-based models are often convenient when there is omission or there are several types of failure (Cox and Oakes, 1984).

Let us formulate some assumptions that are in common use. It is known, from clinical practice, that the risk of failure (hazard of death in cancer) depends on the state of the organism (Manton and Stallard, 1984).

Assumption 14.1
The probability of occurrence of failure that is associated with mortality or morbidity is functionally related to the state of the body. Let the probability of the occurrence of failure at time t for a given trajectory $x_t(w)$, where $w \in \Omega$, be conditional on the path of measurable indices over time:

$$s\left(t|x_t(w)\right) = P\left\{T > t|x_t(w)\right\} \ .$$

Next, as was mentioned above, the dynamics of measurable indices can be consider as a set of realizations of some random process

$$\{x_t(w), \ w \in \Omega, \ t \geq 0\} \ ,$$

that is, defined on the set of observation times $[t_0, t_1, \ldots, T(w)]$. As was discussed in the previous section, the average dynamics of the state variables on the interval $[0, T_{max}]$ is described by the ODE

$$\frac{dx(t)}{dt} = f(x; \alpha), \ x(0) = x_0 \geq 0, \ t \geq 0$$

$$x \in \mathbb{R}^n, \ p \in \mathbb{R}^L \ . \tag{14.4}$$

The individual differences in dynamics are explained by the variation in the model parameters (Marchuk *et al.*, 1989).

Assumption 14.2
For each trajectory $x_t(w_i)$, where $i = 1, \ldots, N$, there exists a piecewise continuous function α_t^i such that

$$x_t(w_i) = x(t, \alpha_t^i), \ t \in [t_0, t_1, \ldots, T_i]$$

where $x(t, \alpha_t^i)$ is a solution of equation (14.4). Then α is replaced by α_t^i.

Assumption 14.3
The unconditional or observed values in the probability of failure within the cohort, $s(t)$, are interpreted as

$$s(t) = E\left\{s\left(t|x_t(w)\right)\right\} \ ,$$

where the conditional probability is averaged over the trajectories of the random process.

A model which functionally relates the survival function to the random process trajectories was suggested by Woodbury and Manton (1977). Specifically, it describes how the distribution of a population of selected state variables changes with time as a function of the following:

- Drift – the deterministic evolution in the mean values of the state variables over time.
- Diffusion – the random changes in individual states due to the influence of endogenous and exogenous factors.
- Homeostatic regression – the change in physiological status, which decreases individual deviation from some reference trajectory of the physiological variables.
- Mortality selection – the removal of persons from the population, conditional on the values of their physiological variants.

In our model we consider the relationship between the deviations of the measurable indices from the reference trajectory and the survival function $s(t)$. A vector α^* determines the solution $x(t, \alpha^*)$ which governs the average dynamics of state variables in the group of patients with favorable disease histories (with maximal life-spans after surgery). The deviations from the reference trajectory $x(t, \alpha^*)$ are caused by unmeasured endogenous and exogenous factors. According to assumption 14.2, this deviation Y_t, given by $Y_t = x_t - x(t, \alpha^*)$, is induced by the fluctuation of the vector p around the value α^*. As a mathematical model for such deviations the SDE was proposed. The random deviations are induced by parameter perturbations and can be considered as a Gaussian–Markovian process which satisfies the linear SDE

$$
\begin{aligned}
dY_t &= a(t)Y_t dt + D(t)dw_t \\
Y_0 &= 0 ,
\end{aligned}
\tag{14.5}
$$

where $a(t)$ is an $m \times m$ matrix, $D(t)$ is an $m \times L$ matrix, and w_t is a Wiener process. The term Y_t is distributed as $N(m(t), \gamma(t))$ where the mathematical expectation and variance satisfy the equations

$$
\begin{aligned}
\frac{d}{dt} m(t) &= a(t)m(t) \\
\frac{d}{dt} \gamma(t) &= a(t)\gamma(t) + \gamma(t)a^T(t) + D(t)D^T(t) .
\end{aligned}
\tag{14.6}
$$

Woodbury and Manton (1977) suggested consideration of the failure rate with respect to the path of physiological covariates over time. Thus,

$$\mu(t, Y_t) = \lim_{\Delta t \to 0} \frac{1}{\Delta t} Pr\left\{T \in (t, t + \Delta t) | T > t, (Y_v, 0 \le v \le t)\right\}$$

is a function which defines the survival chances for an individual with trajectory $(Y_v, 0 \ge v \ge t)$ and

$$s(t|Y_t) = \exp\left(-\int_0^t \mu(u, Y_u) du\right) \ .$$

This function is called the individual mortality rate in Yashin *et al.* (1986). However, this exponential formula does not necessarily hold without some conditions. The necessary and sufficient conditions for this expression were found by Yashin and Arjas (1988). The relationship between the conditional and unconditional failure rates was presented by Yashin (1985). The following proposition is valid.

Proposition 14.1 (Yashin, 1984)
Let $\{Y_t\}$ be an arbitrary, H-adapted random process, and $\mu(Y_t)$ be a non-negative H^Y-adapted function such that for all t we have

$$E\left\{\int_0^t \mu(Y_u) \, du\right\} < \infty \ .$$

Then

$$E\left\{\exp\left(-\int_0^t \mu(Y_u) du\right)\right\} = \exp\left(-\int_0^t E\left\{\mu(Y_u) | T > u\right\} du\right) \ ,$$

where T is a random variable related to the process $\{Y_t\}$ by the formula

$$P\{T > t | H_t^Y\} = \exp\left(-\int_0^t \mu(Y_u) \, du\right)$$

and is interpreted as the moment of death. Here,

$$H_t^Y = \prod_{u > t} \sigma\{Y_v, v \le u\}$$

is a σ-algebra generated by the random process $Y_t(w)$ up to the instant of time t and

$$H^Y = \{H_t^Y\}_{t \ge 0} \ .$$

Hence, the mortality dynamics observed in the group of patients is determined by

$$\lambda(t) = E\{\mu(Y_t) | T > t\} \tag{14.7}$$

where E is averaged over all individual trajectories.

Table 14.2. The patterns of changes in the variance, σ^2, of indices in groups with different types of histories. The term N is the number of measurements.

Index		Aggressive[a]	Torpid[b]	Favorable[c]
EAC-POK	σ^2	7.5409	4.6059	0.3644
	N	75	389	62
E/EAC	σ^2	6.8613	3.2408	0.1879
	N	75	389	60
IgG	σ^2	3.6998	2.7181	0.1295
	N	75	356	75
IgA/IgM	σ^2	6.3628	1.5026	0.0691
	N	75	356	75

[a]Patients who survived less than 12 months after treatment.
[b]Patients survived less than 36 months.
[c]Patients survived more than 36 months.

14.4 Parametrization of the Individual Hazard Function

Unfortunately, there are no general recommendations for the choice of the analytical form of $\lambda(t)$ as a function of measurable variables. The choice of the corresponding analytical relation is determined by the analysis of experimental data and the ease of mathematical manipulations. Alternatively, the simplest approximation of an unknown function is the Taylor series and this may be used to determine the relation. This yields a quadratic failure rate. *Table 14.2* shows the variance of the deviations of measurable indices for three groups of patients with stomach cancer that differ by the pattern of disease. It is important to note that the variance of the deviations of indices from the reference trajectories is inversely proportional to the life-duration of patients:

$$\sigma_1^2 > \sigma_2^2 > \sigma_3^2 \ .$$

Using this fact, the pathological pressure upon the organism due to disease as a function of measurable variables can be approximated by

$$\Lambda(t) = \int_0^t \left(\lambda_0(v) + Y_v^T Q(v) Y_v \right) \, dv$$

where $\lambda_0(t)$ is a function which determines the standard death hazard not related to a given disease, and $Q(t)$ is an $n \times n$ symmetrical, positive, definite matrix. Thus, the conditional survival function in groups of patients can be presented in the form

$$S(t, w) = \exp\left(-\int_0^t \left(\lambda_0(s) + Y_s^T Q Y_s\right) ds\right)$$

The quadratic form of the failure rate function represents increased risk at both high and low physiological values. It specifies a range of values as optimal in the sense of survival. From an alternative viewpoint, when the effect of the factors studied on the failure rate is unknown, it is likely to be non-linear. One simple approximation to the non-linear function is a Taylor series expansion up to the second-order terms. This again yields a quadratic failure rate. The statistical properties of the random process $\{Y_t, t \geq 0\}$ with quadratic failure selection and marginal distribution of the failure time is given by Yashin (1984).

Proposition 14.2 (Yashin, 1984)
Let the random process

$$\{Y_t(w) \in R^n \ , \ t \geq 0 \ , \ w \in \Omega\}$$

satisfy the linear SDE equation (14.5). Also, let the individual failure rate be given by $\mu(t, Y_t) = Y_t^T Q(t) Y_t$ and assume that Y_0 is distributed as $N(m_0, \gamma_0)$ – initial normality. Under these conditions

$$\frac{d}{dt} s(t) = - \left(\lambda_0 + m_t^T Q(t) m_t + tr\left[Q(t)\gamma_t\right]\right) s(t), \quad s(0) = 1 \ ,$$

and Y_t is distributed as $N(m_t, \gamma_t | T > t)$. The terms m_t and γ_t are the solutions of the following ODEs:

$$\frac{d}{dt} m_t = a(t) m_t - 2\gamma_t Q(t) m_t, \quad m(0) = m_0$$

$$\frac{d}{dt} \gamma_t = a(t)\gamma_t + \gamma_t a(t)^T + D(t)(D(t))^T - 2\gamma_t Q(t)\gamma_t,$$

$$\gamma(0) = \gamma_0 \tag{14.8}$$

where

$$m_t = E\{Y_t | T \geq t\}$$

$$\gamma_t = E\left\{(Y_t - m_t)(Y_t - m_t)^T | T \geq t\right\} \ .$$

Since $s(t)$ is the marginal probability of a patient's survival to time t, we will refer to $\lambda(t)$ as the marginal cohort failure rate. The condition given by equation (14.7) is required for inclusion in the subset of survivors alive at time t. Similarly, $N(m_t, \gamma_t | T > t)$ will be called the marginal distribution of Y_t among the survivors at time t. Proposition 14.2 yields the mathematical relationship between the marginal cohort failure rate and the parameters governing change in the mathematical expectations and covariances of the physiological process related to the failure rate. Now the marginal probability density for the random failure time in terms of the statistical properties of the random process is

$$p(t) = \lambda(t) \, \exp\left(-\int_0^t \lambda(v) \, dv\right) \ .$$

Thus, the mortality intensity observed in the group of patients is related to the dynamics of clinically controlled indicators by the formula

$$\lambda(t) = m_t^T Q(t) m_t + tr[Q\gamma_t] \ .$$

Example 14.2

Let us consider the case when the individual trajectories oscillate about some constant level x given by $x = x_0$. Here we have $EY_t = 0$ for all t or $Ex_t = x_0$. At the same time, the variance of the deviations is described by the second equation in (14.6).

So, we can use indicators such as the percentage of lymphocyte for the analysis of disease dynamics. This is a very important fact, because in the majority of cases the measurable indices oscillate about some constant level. This is one reason they were not studied before. However, they have important information about the process in question.

14.5   Estimation of Unknown Parameters

Often the clinical problem leads to a comparison of the failure times in several groups. The crucial point in the application of a survival model is the uncertainty related to unknown parameter values. In general, the model functions $a(t), b(t), Q(t), q(t)$, and $\lambda_0(t)$ are assumed to be known, but the values of the parameters are unknown. In the simplest case the functions a, b, Q, q, and λ_0 are constants. As a rule, the parameters of the model have a practical interpretation. For example, the elements of matrix Q define the degree of influence on the failure dynamics for each state variable. Parameter estimation using clinical data helps to study the role of various factors thought to be related to failure.

14.5.1 Statistical estimations of patient termination time

Consider the Cauchy problem in equation (14.6). According to previous results from Part II, the deviations given by $Y_t = x_t - x(\alpha^*)$ can be approximated by a Gaussian-Markovian process which satisfies the equation

$$dY_t = a(t)Y_t dt + D(t)dw_t, \ Y_0 = 0 \ ,$$

where $DD^T = b\Gamma b^T$.

Under this condition, and with a quadratic mortality rate function given by

$$\Lambda_t = \lambda_0(t) + Y_t^T Q(t) Y_t \ ,$$

we can write

$$
\begin{aligned}
\frac{d}{dt} m_t &= a(t)m_t - 2\gamma_t Q(t)m_t, \ m(0) = m_0 \\
\frac{d}{dt} \gamma_t &= a(t)\gamma_t + \gamma_t a(t)^T + b(t)\Gamma(b(t))^T - 2\gamma_t Q(t)\gamma_t \\
&\quad \gamma(0) = \gamma_0 \ .
\end{aligned}
\tag{14.9}
$$

Assume, for simplicity, that the matrices Q and Γ are diagonal. Introduce a new $2L \times 1$ vector, β, of unknown parameters given by

$$\beta = (\text{diag } Q, \text{diag } \Gamma)^T \ .$$

Let β^* be the true value. For each patient we have a set of measurements:

$$X^i = \left\{ X_{t_k}^i \in \mathbb{R}^n, t_k < T_k, k = 1, \ldots, N_i \right\} \ .$$

Having defined a reference trajectory for a group of N patients, we can obtain the set of individual trajectory deviations given by

$$Y_N = \{Y^i, i = 1, 2, \ldots, N\} \ .$$

Other available observations form the set of failure times in the cohort such that

$$\Theta = \{T_1, \ldots, T_N\} \ .$$

The solution to the Cauchy problem in equation (14.6) can be used to define the unconditional or marginal probability density for random failure times:

$$p(t,\beta) = \lambda(t,\beta) \exp\left[-\Lambda(t,\beta)\right]$$
$$\Lambda(t,\beta) = -\log s(t,\beta)$$

where Λ is the cumulative failure rate function. According to the likelihood principle, we have

$$L(\beta) = \log p(\Theta,\beta) \ ,$$

where

$$p(\Theta,\beta) = \prod_{t\in\Theta} p(t,\beta) \ .$$

In our case,

$$l(\beta) = \sum_{t\in\Theta} \log \lambda(t,\beta) - \Lambda(t,\beta) \ .$$

An estimate of the unknown vector β is given by

$$\beta = \arg \max_{\beta} l(\beta) \ .$$

The difficulty of this estimation procedure is associated with the functions m_t, γ_t, and Λ_t, which are functions of the unknown parameter β.

There are two variants of this estimation problem with respect to the likelihood function.

- We can use the failure times only (see example 14.3).
- We can use the conditional form of the likelihood function. In this case we should consider the random deviations of the process in question. The numerical algorithm for searching such estimates was given by Asachenkov and Sobolev (1993).

Example 14.3
This involves mortality dynamics as a function of observed data. For the patients treated for stomach cancer these estimations have the values

$$Q = \begin{pmatrix} 0.0001 & & & \\ & 0.0074 & & \\ & & 0.0102 & \\ & & & 0.0038 \end{pmatrix} \quad \Gamma = \begin{pmatrix} 0.0324 & & & \\ & 0.0001 & & \\ & & 0.1 & \\ & & & 0.015 \end{pmatrix} .$$

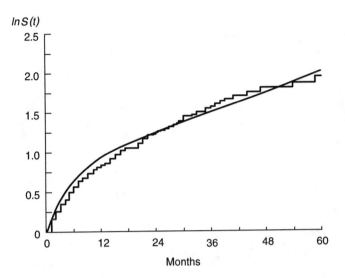

Figure 14.4. The estimation of the survival function. The smooth line is the solution of the model.

These results are based on clinical data from the Research Institute of Oncology in Tomsk. *Figure 14.4* shows the estimates of the survival function and *Figure 14.5* shows the solutions for $m_i(t)$ and $\gamma_i(t)$, where $i = 1, \ldots, 4$.

Note that the mathematical expectation tends to zero at $t \to \infty$, and the variance tends to the level that characterizes the variance in the group of patients with favorable disease histories. The greatest deviations of immunological indices from the reference trajectory, as we can see, occur during the first two years after the beginning of treatment. This result agrees with clinical experience. Now, for estimation of the disease activity of individual patients during treatment we can use an index

$$\mu_t^i = Y_t^T Q Y_t + \lambda_0(t)$$

or

$$M_t^i = \int_0^t \left(Y_u^T Q Y_u + \lambda_0(u) \right) \, du \; ,$$

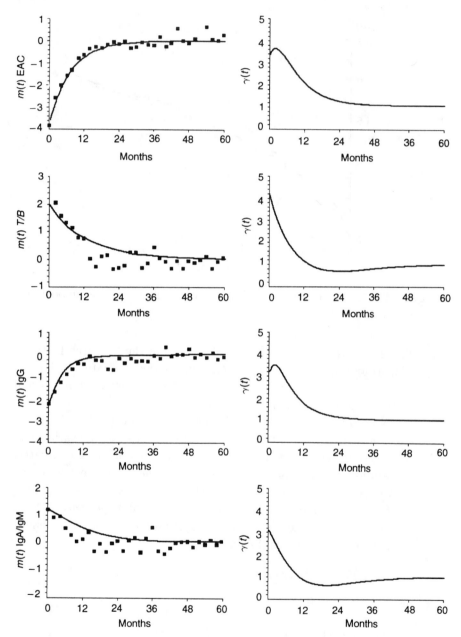

Figure 14.5. Mathematical expectation and variance. Percentage of lymphocytes is not used. The solid line is the solution of the model. The dots are the average deviations from the reference trajectory for a group at the instant of time t.

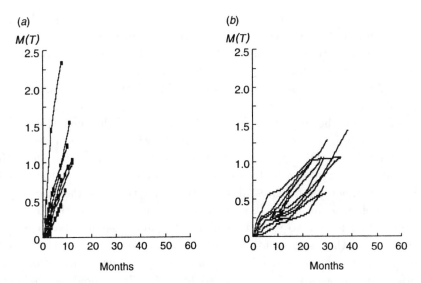

Figure 14.6. Individual estimates of $M(t)$ for two groups of patients: (a) T > 40 months: (b) T < 40 months.

where Y_t is the deviation from the reference trajectory for a particular individual. Since we measure the indicators at discrete instants of time, we can use the approximation

$$M^i(t) = \frac{1}{2} \sum_{k:t_k \leq t} (t_k - t_{k-1}) (\mu^i_{t_k} + \mu^i_{t_{k-1}}) \ .$$

Here, the function $\lambda_0(t)$ and the matrix Q are known. The close correlation between this index and the life-span is shown in *Figure 14.6*.

14.6 Discussion

This estimation can be considered as an integral characteristic of the state of the patient at the instant of time t. *Figure 14.6* gives the assessments of individual mortality intensities, μ_t and M_t, for the patients from different groups which differ in life-duration after surgery. *Figure 14.6* shows that the dynamics of $M(t)$ qualitatively differ for patients with different disease histories. This provides the possibility of checking the development of a clinically non-expressed tumor growth process according to clinically measured indices. It can be used to select an adequate scheme of treatment, taking into account the individual characteristics of the patient. Moreover, the total mortality intensity $M_t(Y(t), Q, \lambda)$ (compare with $\Lambda(t)$ in Section 14.4) as a function of $Y(t)$ [under the known matrix Q and a function $\lambda_0(t)$] can be used to check clinically non-expressed tumor growth on line. We briefly consider this problem in the next chapter.

It should be emphasized that the pattern of the deviation of the measured indices from the reference trajectory is important in the framework of the model suggested. The constructed integral characteristics of the $M(t)$ states can be treated in an analogous manner to the gravity index of infectious diseases and described as the GI of the intensity of the oncological process. It is important to note that in the construction of the model no expert assessments have been used, which is in contrast to the assessments widely applied to constructing the gravity index of the acute form of infectious diseases. The majority of acute infectious diseases are characterized by a significant deviation of clinical indicators at the peak of the disease from their normal values, with their consequent return to norm. In contrast, oncological disease shows irregular deviation of the index values from the reference trajectory that seems to testify to the systematic, stable influence of the tumor on the homeostatic system and in the long term promotes an increase in the probability of death.

The results presented here can be treated as a tool for obtaining a working knowledge about the process in question and as an introduction to the theory of clinical data processing on the basis of available information.

Chapter 15

Further Development of Cancer Research

Even though much progress has occurred in cancer research, many problems are still unresolved. The absence of specific, identifiable differences between normal and malignant cells is a major barrier that has limited the development of specific anti-cancer therapy. Treatment methods have had to rely on the spatial separation of tumor and normal tissues (for surgery or radiotherapy), or on systemic treatments such as chemotherapy. Thus, almost all types of cancer treatments cause significant damage to normal tissue. There is little information on pharmacokinetics and other factors associated with utilizing effective concentrations of each drug at receptor sites: anti-cancer drugs have narrow therapeutic indices, are unstable, and have ill-defined metabolisms.

15.1 The Drug Administration Problem

Systematic treatment with cytotoxic drugs is often the only treatment that may influence all sites of a metastatic disease. At tolerable doses, such treatment can be highly effective for certain types of cancer (lymphomas and testicular cancer), but it is rarely effective for many of the common types of solid tumors (colo-rectal or lung carcinoma). A major limitation to the success of drug treatment is the presence in the tumor of drug-resistant cells which cause either initial resistance to treatment, or subsequent resistance after the tumor has initially responded.

Alternative approaches to cancer treatment include the use of hormones, which may be effective in inducing remission in, but not curing, some cancers that arise in hormone-sensitive tissues such as the breast and the prostate gland. Hyperthermia is being investigated as an alternative or auxiliary treatment which can be combined with ionizing radiation or chemotherapy, but it remains unclear whether heat can provide selective toxicity to tumor cells.

Various approaches to immunotherapy have been tried clinically but classical approaches involving attempted immunization have met with minimal success. The development of monoclonal antibodies and of a number of agents which can stimulate specific components of the immune response (interferons, interleukins, tumor necrosis factor) has rekindled hopes for specific killing of tumor cells via immune-mediated effects.

In this chapter we discuss some aspects of the optimal control problems for oncological patients. As was mentioned in the introduction to Part III, two classes of mathematical models can be considered. The first is the so-called tumor growth model, in which the tumor volume or the number of tumor cells can be calculated by means of direct measurements or by using specific "tumor markers" (see Chapter 11).

Consider the simple example. Let $z(t)$ be the number of tumor cells, where $z(t) > 0$ for $t \geq 0$, and $t = 0$ be the instant of surgery at which the solid tumor is removed. Since radical surgery leads to a decrease in the tumor mass only, and does not affect the cancer mechanism, tumor growth still takes place. Formally, the continuous tumor growth model has the following form:

$$\frac{dz}{dt} = f(z), \quad z(t_0) = z_0 > 0 \ ,$$

where $z(t) \in R_+$ and $f(z)$ is, as a rule, a known non-linear function. If $f(z) = \lambda z \ln(z_{\max}/z)$, where $z_{\max}$ is the maximum size of the tumor and λ is a constant, we have the so-called Gompertz growth equation:

$$\frac{dz}{dt} = \lambda z \ln (z_{\max}/z) \ .$$

An analytical solution of this equation can be obtained by using the transformation $y = \ln(z/z_{\max})$, then

$$\frac{dy}{dt} = -\lambda y \ .$$

Anti-cancer drugs do influence tumor growth. The real problem is to describe the therapy (control) in mathematical terms. If we suggest that the action of a drug is proportional to the population of tumor cells in the form $v(u)z$, where $v(u)$ is a known function, we can write the perturbed tumor growth model in the form

$$\frac{dz}{dt} = \lambda z \ln \left(z_{\max}/z\right) - v(u)z, \quad z(0) = z_0 .$$

Unfortunately, we have no recommendation to help us decide on an appropriate form for $v(u)$. For our example, the following expression looks reasonable (Swan, 1984):

$$v(r) = \frac{k_1 r}{k_2 + r} ,$$

where k_1 and k_2 are constants greater than zero. Now, let $\tau = \lambda t$, $u = r/k_2$, and $p = k_1/\lambda$, then

$$\frac{dy}{d\tau} = -y - \frac{pu}{1+u}, \quad y(0) = -c$$

$$c = \ln \left(z_{\max}/z_0\right), \quad p > c .$$

In Appendix C, Section C.3, we have the same structure for the model which describes the local interactions of tumor cells and effector cells.

Another problem is the selection of suitable performance criteria. One of the simplest ways is as follows. It is known that anti-cancer drugs are toxic. In the absence of additional information, let us consider the following measure:

$$J = \int_0^T u(t) \, dt ,$$

where $u(t)$ is a non-dimensional drug input and t is a non-dimensional time.

Now, using the Pontryagin minimum principle we can write an analytical solution for a control variable:

$$u(\tau) = \left(1 - (c/p)^{1/2}\right)^{-1} \exp \left(\tau/2\right) - 1$$

and

$$y(\tau) = -p \left[1 - \left(1 - (c/p)^{1/2}\right) \exp \left(-\tau/2\right)\right]^2 .$$

Here y tends to $-p$ for large values of time. This means that, even with unbounded control, a plateau level for y is eventually reached beyond which the number of cancer cells cannot be reduced.

The problem becomes more difficult when we deal with models from Chapter 13 or 14. Some of the preliminary results are presented in Sections 15.2 to 15.4, which follow.

15.2   The Controlled Dynamic System

Let us consider the dynamics of indices observed for the patients after
surgery. Let $t = 0$ be the instant of surgery. Let $x(t) \in \mathbb{R}^n$ be a vector
which we can measure using clinical indices. The development of disease
leads to deviations of these indices from the values corresponding to a
healthy organism. Let the dynamics of the clinically measured indices,
on average, be described by the equation

$$\frac{dx(t)}{dt} = f(x; p^*), \quad x(0) = x_0 \geq 0, \quad t \geq 0$$

$$x \in \mathbb{R}^n, \quad p^* \in \mathbb{R}^l .$$

As was mentioned in Chapter 14, the individual trajectories of these
indices presumably are stochastic in character. So, for computer simulations, ODEs with random perturbations of parameters can be used
(see Part II, Chapters 6 and 7):

$$\frac{d}{dt} x_t^\varepsilon = f(x_t^\varepsilon, p^* + \xi_{t/\varepsilon})$$

$$x_0^\varepsilon = x_0, \quad t \in [0, T] .$$

Here, x_t^ε is a perturbed solution, ε (where $\varepsilon > 0$) is a small-valued pa-
rameter, and ξ_t is a stochastic process such that $E\xi_t = 0$ and $\mathrm{cov}(\xi_t, \xi_t + \tau) \to 0$ as $\tau \to \infty$. These trajectories can be considered to be the result
of small perturbations of the dynamic system. The perturbed motion
described by this model is the fast, random fluctuation along the refer-
ence trajectory $x_t(p^*)$. Let Y_t^ε, given by $Y_t^\varepsilon = x_t^\varepsilon - x_t(p^*)$, be a deviation
between the perturbed motion and the reference trajectory. Then, as
was given in Chapter 14, the process Y_t^ε is approximated by the lineal
stochastic differential

$$dY_t = a(t)Y_t dt + b(t)dw_t ,$$

where w_t is a Gaussian stochastic process with independent increments.
    The effect of drug action is manifested by the dynamics of the mea-
surable indices and the survival function. There are two possibilities of
incorporating the control variables in the model:

- The action of an anti-cancer drug is proportional to the population
  of cancer cells.
- Drug administration changes the parameters of the model.

The problem is to construct the functions $a(t, u)$ and $b(t, u)$ on the basis of the available information. Let A_0 be the group of patients who have surgery only and A_1 be the group of patients with surgery and treatment (chemotherapy, etc.). Experimental data from patients with stomach cancer show that the character of the $x(.)$ dynamics in the groups A_0 and A_1 are similar and, consequently, can be described by an equation of the same structure (Asachenkov, 1990). By taking this fact into consideration we can formulate the following hypothesis.

Hypothesis
The dynamics of measurable variables in the groups of patients A_0 and A_1 are described by the same equation with parameters p^* and p_1, where $p^* \neq p_1$, respectively. The difference between the coefficients is determined by the cancer therapy (control):

$$p_1 = p^* + \beta u(t), \quad u(t) \in U, \quad \beta = \text{constant} .$$

As a result of this hypothesis, the perturbed control model is

$$\begin{aligned}
\frac{d}{dt} x_t^\varepsilon(u) &= f(x_t^\varepsilon, p^* + \beta u(t) + \xi_{t/\varepsilon}) \\
x_0^\varepsilon(u) &= x_0(p^*), \quad t \in [0, T] ,
\end{aligned}$$

and the model for the deviations $Z_t = x_t^\varepsilon - x_t(u, p^*)$ can be approximated by the following SDE:

$$\begin{aligned}
dZ_t(u) &= a(t, u)Z_t(u)dt + b(t, u)dw_t \\
Z_0(u) &= 0 .
\end{aligned} \tag{15.1}$$

Here, $EZ_t = 0$ for all $\forall_{t \geq 0} = 0$ and $x_t(u, p^*)$ is a solution of

$$\begin{aligned}
\frac{d}{dt} x_t(u, p^*) &= f(x_t, p^* + \beta u(t)) \\
x_0(u, p^*) &= x_0(p^*), \quad t \in [0, T] .
\end{aligned}$$

Another approach is based on the idea that the parameters of the model are connected by some common factor (see Chapter 10). This means that if p^* is a known vector such that the model

$$\frac{d}{dt} x_t = f(x_t, p^*), \quad x_0 = c, \quad t \in [0, T]$$

$$x_t \in \mathbb{R}^n, \quad p^* \in \mathbb{R}^l$$

describes the average dynamics, then the individual dynamics for the kth patient can be described in the form:

$$\frac{d}{dt} x_t^k = f(x_t^k, p^* + \beta u(t), HL^k), \quad x_0 = c^k, \quad t \in [0, T]$$

$$x_t^k \in \mathbb{R}^n, \quad p^* \in \mathbb{R}^l, \quad HL \in \mathbb{R} \ . \tag{15.2}$$

Thus, for the individual evaluation of the parameters of the model we have to estimate only one parameter, HL, which may be done at the beginning of the treatment for the disease, and then use the model in equation (15.2) for the solution of the optimal drug administration problem.

15.3   A Performance Criterion

In medicine, the following scalar cost functional is sometimes used:

$$J(u) = q(z(T)) + \int_0^T R(z, u) \, dt \ ,$$

where $q(z(T))$ is a function of the state variable at the end of treatment and the second term estimates the cumulative drug toxicity over the time period [0,T]. The optimization problem is to determine the control (administration of drugs) so as to minimize this performance index.

As noted previously in the introduction to part III, there is another way (Swan, 1984). Let $y_r(t)$ be a known function. This function prescribes the "desirable" time course for the measured variable and reflects the experience of the investigator. Define $e(t) = x(t) - y_r(t)$, where $e(t)$ is the deviation between the model-derived value $x(t)$ and its desired level. It seems reasonable to use a drug in such a manner that the measurable variables follow the curve given by $y_r(t)$ with the minimal toxicity effect

$$J(u) = \int_0^T \left(e^T(t) Q(t) e(t) + u^T(t) R u(t) \right) dt, \quad u \in U \ ,$$

where Q and R are weighting matrices.

However, the function $y_r(t)$, as a rule, is determined by examination by experts. Therefore, it contains some elements of subjectivity. Is it possible to choose this function using only objective information? From the previous chapter, the answer is "yes". It may be a reference trajectory. Now (see Chapter 14), it looks reasonable to consider the following cost function:

$$J(T) = E\left\{\int_0^T Y_t^T Q Y_t + u^T R u \, dt\right\} \to \min_u \quad . \tag{15.3}$$

Here, $Y_t(u) = x_t^\varepsilon(u) - x_t(p^*)$. One of the reasons for the appeal of the cost function in the form of equation (15.3) is the following.

Observation
Table 14.2 (Chapter 14) shows the variance of deviations of immunological indices for three groups of patients with stomach cancer with different disease patterns. It is important to note that the variance of deviations of indices from the reference trajectories is inversely proportional to the life duration of patients:

$$\sigma_1^2 > \sigma_2^2 > \sigma_3^2 \quad .$$

The quadratic form [see equation (15.3)] represents increased risk for significant deviations from a reference trajectory. This means that to minimize the pathological pressure upon the organism caused by the disease (tumor growth) is equivalent to maximization of the life-span of the patient after beginning treatment. Thus expression (15.3) could be considered as a good candidate for the cost function.

15.4   Statement of the Problem

A commonly accepted criterion of recovery in oncology is a five-year life-span of a patient after the beginning of his treatment. After this, the pattern of the fall in the number of oncological patients in a group of the same age is determined by the magnitude of the natural mortality factor. This means that we can study the optimal treatment problem on the finite interval only.

Consider the case where $n = 1$ and the performance criterion is of the form

$$J(T) = E\left\{\int_0^T Q Y_t^2 \, dt\right\} \to \min_u \quad .$$

Here, $Y_t(u) = x_t^e(u) - x_t(p^*)$. Using $Z_t = x_t^e(u) - x_t(u, p^*)$ and $EZ_t = 0$ for all $t \geq 0$, we have

$$
\begin{aligned}
E(Y_t^2) &= E\left(Z_t + x_t(u, p^*) - x_t(p^*)\right) \\
&= E(Z_t)^2 + (x_t(u, p^*) - x_t(p^*))^2 \ .
\end{aligned}
$$

For $E(Z_t)^2 = D_t$, using equation (15.1), we have

$$
\frac{d}{dt} D_t = 2a(t, u)D_t + (b(t, u))^2 , \quad D_0 \geq 0 \ .
$$

Now the problem is to minimize the cost function

$$
J(T) = Q\left\{ \int_0^T \left(D_t + (x_t(u, p^*) - x_t(p^*))^2 \right) dt \right\} \to \min_u
$$

under the conditions

$$
\begin{aligned}
\frac{d}{dt} D_t &= 2a(t, u)D_t + (b(t, u))^2 , \quad D_0 \geq 0 \\
\frac{d}{dt} x_t(u, p^*) &= f\left(x_t, p^* + \beta u(t)\right), \quad x_0(u, p^*) = x_0 \\
\frac{d}{dt} x_t(p^*) &= f\left(x_t, p^*\right), \quad x_0(p^*) = x_0 \ .
\end{aligned}
$$

15.4.1 Example: the optimal time for injection

Let $u(t)$ be a known variable, for example,

$$
u(t) = \begin{cases} \text{constant} & \text{for } t \in [\tau^k, \tau_k + \Delta], \quad k = 1, 2, \ldots, N \\ 0 & \text{for } t \notin [\tau_k, \tau_k + \Delta] \end{cases}
$$

$$
\tau_1 \geq \Delta, \quad \tau_1 + (N + 1)\Delta \leq T, \quad \Delta = \text{constant} \ .
$$

Also, we should choose the dose, time, and the number of injections, N, to minimize $J(T)$.

Suppose that the dynamics of the observed index are described by the equation

$$
\frac{d}{dt} x(t, \lambda) = -\lambda x, \quad x(0, \lambda) > 0
$$

$$
x \in \mathbb{R}, \quad \lambda \in \mathbb{R} ,
$$

the support solution has the form

$$
x(t, \lambda) = x_0 \exp\left[-\lambda(t - T_0)\right] \ .
$$

Using the hypothesis from Section 15.2, the dynamics of the observed index with control are described by the equation

$$x(t, \lambda + \beta u(t)) = x_0 \exp\left(-\int_{T_0}^t (\lambda + \beta u(s))\, ds\right) .$$

The equation for the deviation $Z_t = x_t^\varepsilon - x_t(u, p^*)$ has the form

$$dZ_t = -(\lambda + \beta u(t))\, Z_t + \sigma x(t, \lambda + \beta u(t))\, dw_t, \quad Z_0 > 0$$

and

$$\frac{d}{dt} D(t) = -2(\lambda + \beta u(t))\, D(t) + \sigma^2 x^2(t, \lambda + \beta u(t)), \quad D(0) \geq 0 .$$

Let the number of injections N be given by $N = 1$, the treatment interval Δ be given by $\Delta = \text{constant} > 0$, $0 < \Delta T$, $T_0 = 0$, and the dose of drug u be given by $u = 1$ for $t \in [\theta, \theta + \Delta]$ and $u = 0$ in the opposite case. Under these conditions, the simplest optimal-time problem can be formulated as

$$J = \int_0^T \left(D_t + [x(t, \lambda + \beta u) - x(t, \lambda)]^2\right) dt \to \min_\theta .$$

The optimal time θ can be found from the condition

$$J = \int_0^\theta D(t, 0)\, dt + \int_\theta^{\theta+\Delta} D(t, 1)\, dt + \int_{\theta+\Delta}^T D(t, 0)\, dt$$
$$+ \int_\theta^{\theta+\Delta} [x(t, \lambda + \beta u) - x(t, \lambda)]^2\, dt \to \min_\theta .$$

The last integral does not depend on θ. Thus, we can consider the following problem:

$$J = \int_0^\theta D(t, 0)\, dt + \int_\theta^{\theta+\Delta} D(t, 1)\, dt + \int_{\theta+\Delta}^T D(t, 0)\, dt \to \min_\theta .$$

$$(15.4)$$

Introduce a new variable

$$D = z\sigma \exp\left[-(\lambda + \beta u)t\right], \quad \sigma = \sigma^2, \quad \lambda = 2\lambda, \quad \beta = 2\beta ,$$

then

$$\frac{d}{dt} z(t) = -[\lambda + \beta u(t)]\, z(t) + 1, \quad z(0) = D(0)/\sigma \geq 0$$

and equation (15.4) can be rewritten as

$$J = \int_0^T z(t,0)e^{-\lambda t}\, dt + \int_\theta^{\theta+\Delta} z(t,1)e^{-(\lambda+\beta)(t-\theta)}\, dt$$
$$+ \int_{\theta+\Delta}^T z(t,0)e^{-\lambda(t-\theta-\Delta)}\, dt \to \min_\theta \ .$$

Let us denote the terms Q_1, Q_2, and Q_3 as follows:

$$Q_1 = \frac{\partial}{\partial\theta} \int_0^\theta z(t,0)e^{-\lambda t}\, dt$$
$$= a\,(z(0),\lambda)\,e^{-2\lambda\theta} + 1/\lambda\, e^{-\lambda\theta}$$
$$a\,(z(0),\lambda) = z(0) - 1/\lambda\ .$$

$$Q_2 = \frac{\partial}{\partial\theta} \int_\theta^{\theta+\Delta} z(t,1)e^{-\lambda(t-\theta)}\, dt$$
$$= a\,(z(0),\lambda)\, b(\lambda,\beta,\Delta)e^{-2\lambda\theta}$$
$$b(\lambda,\beta,\Delta) = \frac{\lambda}{2(\lambda+\beta)}\,(e^{-2(\lambda+\beta)\Delta} - 1)\ .$$

$$Q_3 = \frac{\partial}{\partial\theta} \int_{\theta+\Delta}^T z(t,0)e^{-\lambda(t-\theta-\Delta)}\, dt$$
$$= e^{\lambda\theta} \left\{ [1/2a(z(0),\lambda)]e^{-(\lambda+\beta)\Delta}e^{-2\lambda(T-\Delta)} + [1/\lambda]e^{-\lambda(T-\Delta)} \right\}$$
$$- e^{2\lambda\theta}e^{-2\lambda(T-\Delta)} \left(1 - e^{-(\lambda+\beta)\Delta}\frac{\beta}{\lambda(\lambda+\beta)} \right)$$
$$- [1/2a(z(0),\lambda)]\, e^{-\lambda\theta}e^{-(\lambda+\beta)\Delta}\ .$$

Now we can calculate the optimal time from the equation

$$Q_1 + Q_2 + Q_3 = 0$$
$$c_1 e^{-\lambda\theta} + c_2 e^{-2\lambda\theta} = q_1 e^{\lambda\theta} + q_2 e^{2\lambda\theta}\ ,$$

where c_1, c_2, q_1, and q_2 are the constants which depend on $z(0)$, λ, β, and Δ. The positive injection time $\theta^* = \theta\,(z(0))$ exists if $c_1 + c_2 > q_1 + q_2$. Also, there is a constant $O\,(z(0))$ such that $\theta \in [0,T]$ if $(c_1 + c_2) - (q_1 + q_2) < O\,(z(0))$.

15.4.2   A more complex example

This example was given by Asachenkov and Grigorenko (1992). Let $u(t)$ be an unknown function and $y(t)$, where $y(t) > 0$, a scalar, denote the state of the system at time t. Also, $y(T_0) = y_0 > 0$, and its dynamical equation is

$$\frac{d}{dt} y = -2\left(\lambda + \beta u(t)\right) y + \left[\sigma_1 + \sigma_2 u(t)\right]^2 x^2(t)$$

$$y, u \in \mathbb{R}^1, \quad t \in [T_0, T], \quad u(t) \in [0, 1]$$

$$x(t) = x_0 \exp\left[-\lambda(T - T_0)\right] , \tag{15.5}$$

where $u(t)$ is a control function chosen so that one minimizes the cost functional:

$$J = \int_{T_0}^{T} \left[Qy(t) + Ru(t)\right] dt \to \min_u . \tag{15.6}$$

Here, $x_0, Q, R, T, \lambda, \beta, \sigma_1$, and σ_2 are positive constants.

Introduce Pontryagin's function for equations (15.5) and (15.6):

$$
\begin{aligned}
H(t, y, \psi, u) &= -(Qy + Ru) + \psi\Big(-2(\lambda + \beta u)y \\
&\quad + (\sigma_1 + \sigma_2 u)^2 x^2\Big) ,
\end{aligned}
\tag{15.7}
$$

where ψ satisfies the adjoint equation

$$\frac{d}{dt} \psi = 2(\lambda + \beta u)\psi + Q \tag{15.8}$$

and the transversality condition $\psi(T) = 0$.

According to Pontryagin's principle, the extremal control $u^*(t, y, \psi)$ satisfies the following conditions: for $\psi < 0$,

$$
u^*(t, y, \psi) =
\begin{cases}
\frac{R}{2\sigma_2^2 x^2 \psi} + \frac{\beta}{\sigma_2^2 x^2} y - \frac{\sigma_1}{\sigma_2} & \text{for } 0 \leq \mu(y, \psi, t) \leq 1 \\
0 & \text{for } \mu(y, \psi, t) < 0 \\
1 & \text{for } \mu(y, \psi, t) > 1
\end{cases}
\tag{15.9}
$$

where

$$\mu(y, \psi, t) = \frac{R}{2\sigma_2^2 x^2 \psi} + \frac{\beta}{\sigma_2^2 x^2} y - \frac{\sigma_1}{\sigma_2} ; \tag{15.10}$$

and for $\psi = 0$,

$$u^*(t, y, \psi) = 0 ; \tag{15.11}$$

and for $\psi > 0$,

$$u^*(t, y, \psi) = \begin{cases} 1 & \text{for} \quad \mu(y, \psi, t) < 0.5 \\ 0 & \text{for} \quad \mu(y, \psi, t) > 0.5 \\ 0 \text{ or } 1 & \text{for} \quad \mu(y, \psi, t) = 0.5 \ . \end{cases} \tag{15.12}$$

Thus we must study the boundary-value problem given in equations (15.5), (15.7), and (15.9)–(15.12) to compute the optimal solution.

Statement 15.1

For the system in equation (15.8), the terminal condition $\psi(T) = 0$ gives a permitted form of control only for the initial values $\psi(T_0)$ from the interval

$$-\frac{Q}{2\lambda} < \psi(T_0) \leq 0 \ , \quad T_0 \in [0, T] \ .$$

Thus, the cases in equations (15.11) and (15.12) can be omitted and we can concentrate our attention on equations (15.5), (15.8), and (15.9). Let us introduce new variables –

$$\begin{aligned} \chi &= y \exp(2\lambda t) \\ \eta &= \psi \exp(-2\lambda t) \end{aligned}$$

– and rewrite our problem in the form

$$\begin{aligned} \frac{d}{dt}\chi &= -2\beta u^*\chi + (\sigma_3 + \sigma_4 u^*)^2, \quad \chi(T_0) = y_0 \exp(2\lambda T_0) \\ \frac{d}{dt}\eta &= 2\beta u^*\eta + Q \exp(-2\lambda t), \quad \eta(T) = 0 \end{aligned} \tag{15.13}$$

$$\begin{aligned} u^*(\chi, \eta, t) &= \nu(\chi, \eta) \text{ if } 0 \leq \nu(\chi, \eta) \leq 1 \\ u^*(\chi, \eta, t) &= 1 \text{ if } \nu(\chi, \eta) > 1 \\ u^*(\chi, \eta, t) &= 0 \text{ if } \nu(\chi, \eta) < 0 \ . \end{aligned}$$

Here,

$$\begin{aligned} \nu(\chi, \eta) &= \frac{R}{2\sigma_4^2\eta} + \frac{\beta}{\sigma_4^2}\chi - \frac{\sigma_3}{\sigma_4} \\ \sigma_3 &= \sigma_1 x_0 \exp(\lambda T_0) \\ \sigma_4 &= \sigma_2 x_0 \exp(\lambda T_0) \ . \end{aligned}$$

For $u = 0$ the system in equation (15.13) has the form

$$\frac{d}{dt}\chi = \sigma_3^2$$

$$\frac{d}{dt}\eta = Q\exp(-2\lambda t), \quad \eta(T) = 0 . \tag{15.14}$$

Then, we make the following statement.

Statement 15.2
The solution of the system in equation (15.14) on the interval $[0, T]$ satisfies the initial condition $\eta(T) = 0$ only for

$$\frac{-Q\left[1 - \exp\left(-2\lambda T\right)\right]}{2\lambda} \le \eta(0) \le 0 .$$

The following conditions hold for the trajectories $\chi(\eta) : \chi'_\eta > 0$ for all $\eta < 0$; $\chi''_\eta > 0$ if $-Q/2\lambda < \eta < 0$; and $\chi''_\eta < 0$ if $\eta < -Q/2\lambda$. The phase portrait of the system in equation (15.14) is presented in *Figure 15.1*.

Consider the system in equation (15.13) for $u = 1$. This is also non-stationary and has the form

$$\frac{d}{dt}\chi = -2\beta\chi + (\sigma_3 + \sigma_4)^2, \quad \chi(T_0) = y_0\exp(2\lambda T_0)$$

$$\frac{d}{dt}\eta = 2\beta\eta + Q\exp(-2\lambda t) . \tag{15.15}$$

The phase portrait of the system is presented in *Figure 15.2*.

Now we study the system in equation (15.13) for $u = \nu(\chi, \eta)$. We have

$$\frac{d}{dt}\chi = \frac{R^2}{4\sigma_4^2\eta^2} - \frac{\beta^2}{\sigma_4^2}\chi^2 + \frac{2\beta\sigma_3}{\sigma_4}\chi$$

$$\frac{d}{dt}\eta = \frac{2\beta^2\eta\chi}{\sigma_4^2} - \frac{2\beta\sigma_3\eta}{\sigma_4} + Q\exp\left(-2\lambda t\right) + \frac{R\beta}{\sigma_4^2} . \tag{15.16}$$

In the domain $0 \le \nu(\chi, \eta) \le 1$, we have

$$-\frac{R}{2\beta\eta} + \frac{\sigma_3\sigma_4}{\beta} \le \chi \le -\frac{R}{2\beta\eta} + \frac{(\sigma_3 + \sigma_4)\sigma_4}{\beta} . \tag{15.17}$$

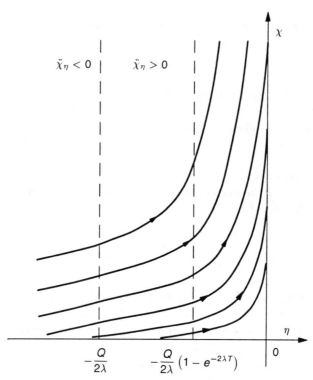

Figure 15.1. The phase portrait of the system in equation (15.14).

To study the system in equation (15.16), introduce four line constants as follows:

$$\Gamma = \left\{ (\chi, \eta) : \chi = -\left(\frac{Q\sigma_4^2 \exp(-2\lambda t)}{2\beta^2} + \frac{R}{2\beta} \right) \frac{1}{\eta} + \frac{\sigma_3\sigma_4}{\beta}, \eta < 0 \right\} \quad .$$

This is a set of points for which $\eta_t' = 0$. Then,

$$\Pi = \left\{ (\chi, \eta) : \chi = \frac{1}{\beta} \left[\sigma_3\sigma_4 + \left(\sigma_3^2\sigma_4^2 + \frac{R^2}{4\eta^2} \right)^{1/2} \right], \ \eta < 0 \right\} ,$$

which is a set of points for which $\chi_t' = 0$. Next,

$$H = \left\{ (\chi, \eta) : \chi = -\frac{R}{2\beta} \frac{1}{\eta} + \frac{\sigma_3\sigma_4}{\beta}, \ \eta < 0 \right\} ,$$

which is a lower boundary of the domain in equation (15.17). Finally,

$$B = \left\{ (\chi, \eta) : \chi = -\frac{R}{2\beta} \frac{1}{\eta} + \frac{(\sigma_3 + \sigma_4)\sigma_4}{\beta}, \ \eta < 0 \right\} ,$$

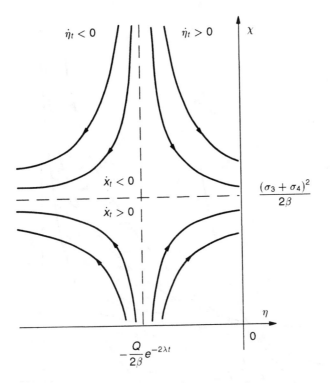

Figure 15.2. The phase portrait of the system for $u = 1$.

which is an upper boundary for equation (15.17).

The lines II, Γ, and B in *Figure 15.3* are situated above the line H in the plane (η, χ). The abscissa of the intersection point for II and B is

$$\eta = -\frac{R}{(\sigma_3^2 - \sigma_4^2)} \ .$$

One of the intersection points for Γ and B is

$$\eta = -\frac{Q \exp(-2\lambda t)}{2\beta} \ .$$

Let C be the point of intersection for II and Γ. It is a moving saddle point (Bylov *et al.*, 1966). The form of separatries are given in *Figure 15.3*. The abscissa of the point C is

$$\eta = -\frac{1}{2\sigma_3 \exp(2\lambda t)} \left(\frac{Q^2 \sigma_4^2}{\beta^2} + \frac{2Q \exp(2\lambda t)}{\beta} \right)^{1/2} \ .$$

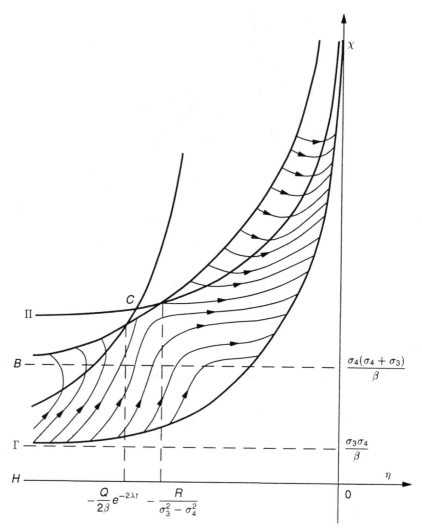

Figure 15.3. The form of separatries and phase trajectories for $Q/2\beta > R/(\sigma_3^2 + \sigma_4^2)$.

By virtue of the fact that the system is non-stationary, the point C moves from left to right. If $Q/2\beta > R/(\sigma_3^2 + \sigma_4^2)$, then C is outside of the domain in equation (15.17) for $0 < t \leq \ln \left[Q(\sigma_3^2 - \sigma_4^2)/2\beta R\right]/2\lambda$. The associated phase trajectories are given in *Figure 15.3*. For $t > \ln \left[Q(\sigma_3^2 - \sigma_4^2)/2\beta R\right]/2\lambda$, the point C moves inside a domain. If $Q/2\beta <$

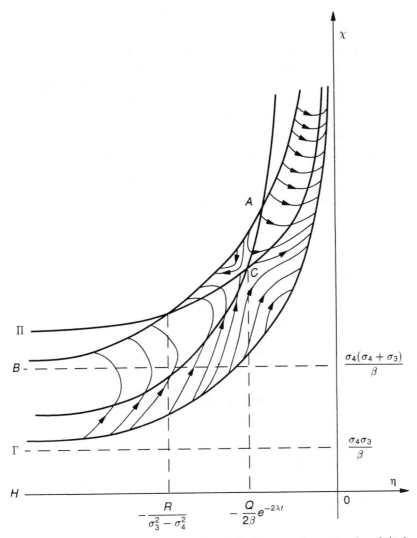

Figure 15.4. The form of separatries and phase trajectories for $Q/2\beta < R/(\sigma_3^2 + \sigma_4^2)$.

$R/(\sigma_3^2 + \sigma_4^2)$, then C is inside of the domain given in equation (15.17) for $t > 0$. The associated phase trajectories are given in *Figure 15.4*.

Now consider the problem of synthesis. The conditions of the Filipov theorem (Warga, 1972) are realized for the extremal problem in equations (15.5) and (15.6). This means that for this problem the optimal trajectory exists for any initial state D_0 and those values of T_0 and

T for which the controllability problem on the terminal set is solvable. Moreover, for this system the conditions of existence, uniqueness, and non-local continuity are realized. According to the maximum principle, the lines B and H switch over. The control conditions are $u = 1$ above the line B, $u = \nu(\eta, \chi)$ between B and H, and $u = 0$ below the line H.

Let T_3 be the time of transfer of the system, using $u = 0$, from an initial state on line H at $t = T - T_3$ to the state $\chi(T)$, $\eta(T) = 0$ at $t = T$. We can calculate T_3, using equations (15.14) and (15.17) and the definition of T_3, by means of the transcendental equation

$$Q\beta\chi(T) - Q\beta\sigma_3^2 T_3 = Q\sigma_3\sigma_4 + \frac{R\lambda \exp(2\lambda T)}{\exp(2\lambda T_3) - 1} \ . \tag{15.18}$$

Statement 15.3

A solution of equation (15.18) for T_3 exists:

$$
\begin{aligned}
T_3^* &= \frac{1}{\lambda} \ln \left[\frac{1}{2A^{1/2}} + \left(\frac{1}{4A} + 1 \right)^{1/2} \right] \\
A &= \frac{\sigma_3^2 \beta Q}{2R\lambda^2 \exp(2\lambda T)} \ .
\end{aligned}
\tag{15.19}
$$

This takes place, if the following inequality holds:

$$B > 2, \quad \tau > -\frac{1}{2\lambda} \ln \left(1 - \frac{2}{B} \right)$$

$$B = \frac{2\sigma_3}{\lambda} \left(\frac{Q\beta}{2R} \right)^{1/2} \ ,$$

then $T_3 < T$. If equation (15.18) is realized, and if $\chi(T) \to \infty$, then $T_3 \to 0$ for all fixed T.

Let T_2 be the time of transfer of the system, using $u = \nu(\eta, \chi)$, from the state on line B at $t = T - T_3 - T_2$ to the state on line H at $t = T - T_3$. From the system in equation (15.16), it is clear that T_2 is finite only for the initial values situated above point A (see *Figure 15.4*). If we fix $\chi(T)$ and integrate the system in equation (15.13) in inverse time, we can calculate the values $\chi(0)$ and $\eta(0)$ (see *Figure 15.5*).

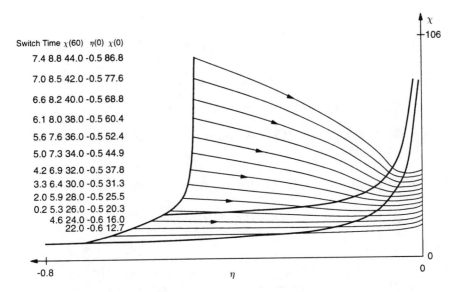

Figure 15.5. The solution of the synthesis problem.

15.4.3 Discussion

Unfortunately, these examples (Sections 15.4.1 and 15.4.2) have some defects. First, the individual characteristics of patients are not taken into account. The second defect is related to the toxicity problem.

Individual Therapy

There are some ways to take into account the individual characteristics of a patient. First, we can use individual initial values in the model. Second, a more sophisticated method, is to use individual coefficients. Consider the first possibility. Divide the interval $[0, T]$ in two: $[0, T_0)$ and $[T_0, T]$. Then,

$$J(T) = J(T_0) + J(T - T_0) \ .$$

Let $Y_i(u, t)$ be a known individual realization on $[0, T_0)$. Consequently, we can estimate $J(T_0)$ and $x_t(u, p^*)$, $x_t(p^*)$ for $t = T_0$. Let $J^*(T_0)$ be an estimate of the cost function on the interval $[0, T_0)$, which can be made as follows:

$$J^*(T_0) \approx 0.5Q \sum_{t_k, t_k \leq T_0} (t_k - t_{k-1}) \left(Y_i^2(t_k) + Y_i^2(t_{k-1}) \right) \ .$$

Now we need to estimate $D(T_0)$. Using the equation for D_t, we can formally write

$$
\begin{aligned}
D(t, u) &= \exp\left(2\int_0^t a(t, u)\, dt\right)\left[D(0) + \int_0^t [b(\tau, u)]^2\right. \\
&\qquad \left.\times \exp\left(-2\int_0^\tau a(s, u)\, ds\right)\right] d\tau \\
&= D(0)c_1(t, u) + c_2(t, u), \quad t \le T_0 \; .
\end{aligned}
$$

Here, $D(0)$ is unknown. To make an estimation of this value possible, we use the relation

$$
\begin{aligned}
J^* \approx J(T_0) &= Q\int_0^{T_0}\Big(D(0)c_1(t, u) + c_2(t, u) \\
&\qquad + [x_t(u, p^*) - x_t(p^*)]^2\Big)\, dt \; .
\end{aligned}
$$

Then,

$$
D^*(0) \approx \frac{J^*(T_0) - Q\int_0^{T_0}\Big(c_2(t, u) + [x_t(u, p^*) - x_t(p^*)]^2\Big)\, dt}{Q\int_0^{T_0} c_1(t, u)\, dt} \ge 0 \; .
$$

Now, $D(T_0) = D^*(0)c_1(T_0, u) + c_2(T_0, u)$: thus, the drug administration problem for a particular individual can be formulated as follows:

$$
J(T - T_0) = Q\left\{\int_{T_0}^T\Big(D_t + [x_t(u, p^*) - x_t(p^*)]^2\Big)\, dt\right\} \to \min_u \; .
$$

Here, the disease history of the individual is taken into consideration by means of the initial conditions in the equation for variance.

Toxicity

This is the major problem with anti-cancer drugs. A number of clinicians suggest that toxic effects can be characterized by the integral

$$
\text{toxicity} \propto \int_{t \in T_u} u(t)\, dt \; ,
$$

where $u(t)$ is the drug concentration in the plasma. Now the cost function can be written, for example, in the form

$$
\begin{aligned}
J(T - T_0) &= \int_{T_0}^T Q\Big(D(t, u) + [x_t(u, p^*) - x_t(p^*)]^2\Big)\, dt \\
&\quad + \int_{T_0}^T Ru(t)\, dt \to \min_u \; .
\end{aligned}
$$

In the previous sections, we give preliminary results for the optimal therapy problem (based on observations) for patients after radical surgery. The results presented here can be treated as a tool for obtaining a working knowledge about the process in question and as an introduction to the mathematical theory of optimal drug administration taking into account the individual characteristics of patients.

15.5 Directions for Future Research

The nature of the immune system is typical of the non-linear systems which are prevalent throughout socio-economics and environmental processes. Consequently, the methods of analysis, decision making, and the related software which are developed for one system may be just as valid for the others. For example, the chemical law of mass action is just as basic to environmental pollution as to immunology. Moreover, the resulting mathematical models have the same structure. Such parametric control structure also appears in economics, for example, in the form of tax rates and investment ratios. Cellular diffusion in the immune system is typical of demographic compartmental migration of populations, and both are similar to chemical particle diffusion, whether in immunology or ecology. Obviously, results such as the estimation of model parameters, estimations or observations of state, stability, control, modeling, and simulation can be applied to these multidisciplinary processes.

Mathematical models of disease and methods for processing clinical data provide a foundation for more applications-oriented research related to multidisciplinary problems concerning the environment, as well as clinical practice. One of the many results of these studies is the possibility of estimating human health levels from simple, laboratory tests (Chapter 10). These studies could be developed to cover areas such as demographical data analysis. This enables us to move from the organism level to the population level.

Analysis of demographical indices, based on the study of processes at the organism level, allows us to find reasons for the fact that changes in the indices are dependent on social-economical conditions and environmental factors. This study now enables us to monitor the health level of the population in a given region. The result may be a map of regional health levels. This map could be considered as a basis for decision making on the regional or the international level concerning the distribution of social-economical help, environmental policies, and medical prophylaxis methods for the given region. Looking at the health-level map over some time interval, we can estimate the efficiency of these policies. This problem is very important for such regions as Chernobyl, for example, as well as many others in the world.

These studies require substantive knowledge of biomathematics, population dynamics, economics, and the environment; data analysis requires the application of mathematics and computers. Medical or health-oriented research naturally depends on experimental data. The question of data availability becomes a crucial one. Clinical, demographical, and experimental data are needed. Clinical data include parameters of blood, immunity, etc., which may be measured as dynamic values of a state vector of a mathematical model for healthy, as well as infected, people.

Routine, health-related, demographical statistical data, such as the World Health Organization (WHO) provides on cause-specific mortality for all countries, are collected by official statistical agencies and sent on to the WHO headquarters in Geneva, either on tapes or diskettes. They are necessary for finding the connection between demographical indices and health level and for explaining observed values of demographical indices.

Environmental data are needed to study the effects of contaminants on human health. For example, some relationship between air pollution and lung cancer is usually assumed but not methodically derived.

The solution of such a complex problem is possible only with cooperation at the international level to provide independent and objective analyses together with unconstrained interpretation. Such expertise may be very useful for government decision making at all levels.

Appendix C

Other Models in Cancer Research

The interest in cancer models was induced by the necessity of calculating the allowable level of carcinogens in the environment (some chemical materials, radiation, etc.) and of estimating the consequences of exceeding these levels. These carcinogenesis models study the process of the transformation of normal cells into malignant cells and tumor growth.

C.1 Carcinogenesis Models

Models of carcinogenesis study the connection between the frequency and the period of carcinogen exposition or dose rate and the parameters which define the sensitivity of the organism to a carcinogen. Theoretical carcinogenesis deals with the ways in which malignant disease comes into being. Its scope ranges from detailed, molecular theories of the nature of cancer to phenomenological descriptions of the probability of tumor occurrence under specific conditions. The phenomenological end of the spectrum provides the less ambitious goals and also lends itself more readily to the application of mathematical models.

These models are based on the analysis of time intervals $T = \{T_j, j = 1, \ldots, M, 0 \leq T_j < \infty\}$ from the beginning of the transformation of normal cells into malignant ones until the instant of time when the tumor is detected. Here, M is the number of patients in the group. The T_j terms, where $T_j > 0$, $j = 1, \ldots, M$, are known time intervals.

C.1.1   Preliminary notes

Let the time from the beginning of the transformation of the normal cell into a malignant cell until the time when the tumor is detected have a random value given by

$$T = t_1 + t_2 \ ,$$

where t_1 is the transformation time and t_2 is the tumor growth time.

Assume that t_1 and t_2 are independent, random values with distribution functions $G(t)$ and $F(t)$ accordingly. Then the distribution function of T has the form

$$R(t) = Pr\{T < t\} = \int_0^t G(t - \tau) dF(\tau) \ .$$

If the probability density function exists, then

$$r(t) = \int_0^t g(t - \tau) f(\tau) \, d\tau = \int_0^t g(\tau) f(t - \tau) \, d\tau \ .$$

Denote the risk or rate of the occurrence of new tumors at the instant of time t by

$$
\begin{aligned}
h(t) &= \lim_{\Delta t \downarrow 0} \frac{1}{\Delta t} Pr\left\{T \in [t, t + \Delta t) \mid T > t\right\} \\
&= \frac{d}{dt} Pr\{T < t\} / Pr\{T \geq t\} = \frac{r(t)}{1 - R(t)} \ .
\end{aligned}
$$

Then

$$Pr\{T < t\} = 1 - \exp\left(-\int_0^t h(s) \, ds\right) \ .$$

Example: exponential distribution
If $h(t) = h > 0$, then

$$Pr\{T < t\} = 1 - e^{-ht}, \quad t > 0 \ ,$$

with the mathematical expectation $E\{T\} = 1/h$ and the variance $D\{T\} = 1/h^2$. We can obtain such a distribution in the following way.

Consider the set of cells, their number being given by N, where $N > 1$, which contains identical, independent, normal cells, each of which, with probability $R(t)$, becomes a malignant cell on the interval $(0, t)$. Then the probability $P_j(t, N)$ that the population from N cells will give j malignant cells on the interval $(0, t)$ is described by the binomial distribution

$$P_j(t, N) = \frac{N!}{j!(N-j)!} \, [R(t)]^j \, [1 - R(t)]^{N-j}$$

$$j = 0, 1, \ldots, N \ .$$

For $N \gg 1$ and $R \ll 1$, we have

$$P_j(t, N) \sim \exp[-NR(t)] \, \frac{[NR(t)]^j}{j!} \ . \tag{C.1}$$

Using (C.1) we can calculate the probability that N cells will be given at least one tumor on $(0, t)$:

$$Pr\{T < t\} = 1 - P_0(t, N) = 1 - \exp[-NR(t)] \ .$$

Example: gamma distribution

Consider the particular case of a gamma function when c is an integer, $t > 0$, $\mu > 0$:

$$h(t) = \frac{\mu(\mu t)^{c-1} e^{-\mu t}}{(c-1)! \, [1 - I(c, \mu t)]} \ ,$$

where $I(c, \mu t)$ is an incomplete gamma function of the form

$$I(c, x) = \frac{1}{(c-1)!} \int_0^x u^{c-1} e^{-u} \, du$$

$$= 1 - \sum_{i=0}^{c-1} \frac{x^i e^{-x}}{i!} \ .$$

If $c = 1$, then $h(t) = \mu$ and

$$Pr\{T < t\} = \int_0^t \frac{\mu(\mu t)^{c-1} e^{-\mu t}}{(c-1)!} \, ds, \ t > 0 \ .$$

For the integer c,

$$Pr\{T < t\} = 1 - \sum_{i=0}^{c-1} \frac{\mu t^i e^{-\mu t}}{i!}, \ t > 0 \ .$$

The mathematical expectation and variance are $E\{T\} = c/\mu$, $D\{T\} = c/\mu^2$. See *Figure C.1.(a)*.

The gamma distribution with integer parameter of shape can be considered as the distribution of the sum of the random variables with the same exponential distribution. In biology, the hazard function which takes the form of a gamma function arises in the case when we have a sequence of reactions with the same constant rate μ.

Example: Weibull's distribution

In this case the hazard function has the form

$$h(t) = \mu\beta(\mu t)^{\beta-1}, \quad t > 0, \quad \mu, \beta > 0 \ .$$

If $\beta = 1$, then $h(t) = \mu$ and

$$Pr\{T < t\} = 1 - \exp\left(-(\mu t)^\beta\right), \quad t > 0 \ .$$

The formulae for mathematical expectation and variance have the following forms:

$$
\begin{aligned}
E\{T\} &= \frac{1}{\mu}\Gamma\left(1+\frac{1}{\beta}\right) \\
D\{T\} &= \frac{1}{\mu^2}\left[\Gamma\left(1+\frac{2}{\beta}\right) - \Gamma^2\left(1+\frac{1}{\beta}\right)\right] \ ,
\end{aligned}
$$

where

$$\Gamma(x) = \int_0^\infty u^x e^{-u} \, dx \ .$$

This distribution is named after Waloddi Weibull, a Swedish physicist, who used it in 1939 to represent the distribution of the breaking strength of materials. In the Russian literature it is sometimes called the Weibull–Gnedenko distribution. Usually, there is no explicit theoretical reasoning indicating that a Weibull distribution should be used – it is just that a power transformation is a practical, convenient way of introducing some flexibility in the model.

C.1.2 Transformation models

Consider some simple models. The main aim is to construct the density function $g(t)$.

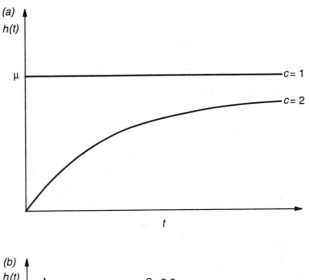

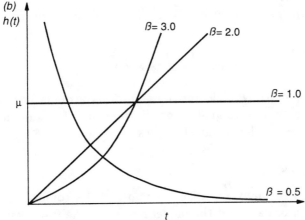

Figure C.1. (*a*) Gamma hazard function. (*b*) Weibull's hazard function.

One-Step Model

One of the first models was suggested by Iverson and Arley (1950). Two cell states are considered: a "0" indicates normal cells and a "1" indicates malignant cells (*Figure C.2*). Let $P_j(t)$, where $j = 0, 1$, be the probability that at the instant of time t a cell is in a state with number j. The term $\lambda(t)$ is the rate of transformation. The model equations are

$$\frac{dP_0(t)}{dt} = -\lambda(t)P_0(t), \quad P_0(0) = 1$$

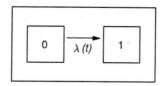

Figure C.2. One-step model.

$$\frac{dP_1(t)}{dt} = \lambda(t)P_0(t), \quad P_1(0) = 0 \ . \tag{C.2}$$

The solution of equation (C.2) has the following form:

$$P_0(t) = \exp\left(-\int_0^t \lambda(s)\,ds\right)$$

$$P_1(t) = 1 - \exp\left(-\int_0^t \lambda(s)\,ds\right) \ .$$

Note, that $P_0 + P_1 = 1$ for all $t \geq 0$. The probability density function $g(t)$ is

$$g(t) = \frac{dP_1(t)}{dt} = \lambda(t)\exp\left(-\int_0^t \lambda(s)\,ds\right) \ .$$

The Iverson–Arley model provides a prototype of mathematical theories that can be used to analyze or interpret data from animal experiments or human epidemiological studies.

Multistage Model

The attempts of many scientists to improve the conformity between the experimental data and the model output in the framework of such a transformation scheme led to the construction of a multistage model. The modifications include: a normal cell goes through a number of stages before it becomes a malignant cell, the consideration of losing cells, and the initial number of cells which are needed to form a tumor. The typical multistage model can be described as follows. Let $P_i(t)$ be the probability that the cell is in stage i, where $i = 0, 1, \ldots, k - 1$, let $\lambda_i(t)$ be the rate of conversion from $i - 1$ to i, and let $\mu_i(t)$ be the rate of death (*Figure C.3*). So, we have

$$\frac{dP_0(t)}{dt} = -[\lambda_0(t) + \mu_0(t)]\,P_0(t), \quad P_0(t) = 1$$

$$\frac{dP_i(t)}{dt} = -[\lambda_i(t) + \mu_i(t)]\,P_i(t) + \lambda_{i-1}(t)P_{i-1}(t)$$

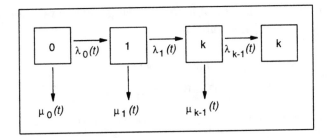

Figure C.3. Multistage model.

$$i = 1, \ldots, k - 1, \quad P_i(0) = 0$$

$$\frac{dP_k(t)}{dt} = \lambda_{k-1} P_{k-1}(t), \quad P_k(t) = 0 \ . \tag{C.3}$$

Define the following terms:

$$L_i(t) = \int_0^t [\lambda_i(s) + \mu_i(s)] \, ds, \quad i = 0, 1, \ldots, k - 1$$

$$L_k(t) = 0 \ .$$

The solution of equation (C.3) has the form

$$P_0(t) = e^{-L_0(t)}$$

$$P_i(t) = e^{-L_i(t)} \int_0^t \lambda_{i-1}(s) P_{i-1}(s) e^{L_i(s)} \, ds, \quad i = 1, \ldots, k \ .$$

Now we can find the probability density function

$$g(t) = \lambda_{k-1}(t) P_{k-1}(t) \ .$$

Multistage Model with Age Structure

Let a be the age of cells in the ith stage. The age is calculated from the first cell in the ith stage. The probability that the cell s is in the ith stage is a function of time and age denoted by $P_i(t, a)$, where $i = 0, 1, \ldots, k$. We can then write the equation

$$\frac{\partial P_i(t, a)}{\partial t} + \frac{\partial P_i(t, a)}{\partial a} = -[\lambda_i(t, a) + \mu_i(t, a)] \, P_i(a, t)$$

$$i = 0, 1, \ldots, k - 1$$

$$P_i(t, 0) = \int_0^\infty \lambda_{i-1}(t, a) P_{i-1}(t, a) \, da, \quad i = 1, \ldots, k \ .$$

Suppose that the age of a normal cell is not relevant for this model, and initially all cells are normal with

$$\frac{P_0(t)}{dt} = -[\lambda_0(t) + \mu_0(t)]\, P_0(t)$$
$$P_0(t) = 1, \quad P_i(0,a) = 0, \quad i = 1,\ldots,k \ .$$

The probability density function of transformation time is

$$g(t) = \int_0^\infty \lambda_{k-1}(t,a) P_{k-1}(t,a)\, da \ .$$

The multistage theory was developed by Muller (1951), Nordling (1954), Stocks (1953), and Armitage and Doll (1954). The central assumption of the multistage theory is that a single clonogenic cell is transformed into a malignant cell as a result of experiencing a sequence of k events, none of which can bring about transformation on its own. These may be thought of as being a sequence of k somatic mutations affecting DNA, although this interpretation is not the only one. An epigenetic version of the multistage theory has been developed by Watson (1977). This involves gene-switching networks, rather than point mutations, and was invoked to make the mathematical model consistent with the biological idea that malignant change might involve a derangement of the cellular differentiated state, rather than a series of conventional mutations.

Another interesting use of the multistage theory was in relation to experiments reported by Peto *et al.* on the role of aging in cancer induction in mice (Peto *et al.*, 1975). An additional extension of the multistage model has been suggested by Whittemore and Keller (1978). Their analysis has been further developed by Klawansky and Fox who have applied it to data on cancer of the lung and of the breast (Klawansky and Fox, 1984).

C.1.3   Parametrization

The probability density functions of transformation and growth can be described in terms of transition rates, for example, $\lambda(t)$. It is impossible to measure these functions directly. In the cases of practical interest we should parametrize these functions using measured indicators. Unfortunately, a regular method for parametrizing does not exist. Let us consider some examples.

The rate $\lambda(t)$ at which normal cells are transformed into a malignant state may be given by

$$\lambda(t) = s + pC(t) \, ,$$

where s is the probability of spontaneous transformation, $C(t)$ is the time dependence of the concentration of the carcinogen, and p is a proportionality constant expressing the dose–response relationship for the carcinogen. Such a model can be used to make predictions about tumor incidence for a variety of assumptions regarding $C(t)$.

In another example, the rate may be

$$\lambda(t) = \lambda_0(t) \, \exp\left(\sum_{j=1}^{m} \beta_j z_j \right)$$

Here, $\lambda_0(t)$ is the normal or basic rate, for which we have not investigated any factors, the z_j terms, where $j = 1, \ldots, m$, are measurements, and β_j are weighting coefficients.

C.1.4 The growth models

Consider the classical model of growth. Our aim is to construct the probability density function of time growth, $\gamma(t)$. Let $\{x(t) > 0; 0 \leq t < \infty\}$ be a set of integer, random variables $0, 1, 2, \ldots$. Consider the Markovian process with the stationary transition probability $P_{ij}(t)$:

$$
\begin{aligned}
P_{i,i+1}(h) &= \lambda_i h + o(h), \quad h \to 0, \ i \geq 0 \\
P_{i,i-1}(h) &= \mu_i h + o(h), \quad h \to 0, \ i \geq 1 \\
P_{ii}(h) &= 1 - (\lambda_i + \mu_i)h + o(h), \quad h \to 0, \ i \geq 0 \\
P_{ij}(0) &= \delta_{ij} = \begin{cases} 1, & i = j \\ 0, & i \neq j \end{cases} \\
\mu_0 &= 0, \ \lambda_0 > 0, \ \mu_i, \lambda_i > 0, \ i = 1, 2, \ldots \ .
\end{aligned}
$$

The parameters λ_i and μ_i, where $i = 1, 2, \ldots$, are called the infinitesimal rate of birth and death, respectively.

Assuming that the probability increase or decrease in the number of cells is proportional to h at $h \to 0$, to calculate the probability that at the instant of time t the population consists of n cells, we need to know the probability distribution at $t = 0$. If $q_i = Pr\{x(0) = i\}$, then

$$Pr\{x(t) = n\} = \sum_{i=0}^{\infty} q_i P_{in}(t) \ .$$

For the transition probability (Karlin and Taylor, 1975) we can write two systems of equations:

$$\frac{dP_{0j}(t)}{dt} = \lambda_0 P_{0j}(t) + \mu_0 P_{1j}(t)$$

$$\frac{dP_{ij}(t)}{dt} = \mu_i P_{i-1,j}(t) - (\lambda_i + \mu_i)P_{ij}(t) + \lambda_i P_{i+1,j}(t)$$

$$P_{ij} = \delta_{ij}, \quad i \geq 1 .$$

This system is known as Kolmogorov's inverse differential equation. However, the system

$$\frac{dP_{i0}(t)}{dt} = -\lambda_0 P_{i0}(t) + \mu_1 P_{i1}(t)$$

$$\frac{dP_{ij}(t)}{dt} = \mu_{j+1} P_{i,j+1}(t) - (\lambda_j + \mu_j)P_{ij}(t) + \lambda_{j-1} P_{i,j-1}(t)$$

$$P_{ij} = \delta_{ij}, \quad j \geq 1$$

is known as Kolmogorov's direct equation. If $\lambda_i = \lambda$ and $\mu_i = \mu$, then the probability that at the instant of time t we have n cells can be described by

$$\frac{dP_n}{dt} = \lambda(n-1)P_{n-1}(t) - (\lambda + \mu)P_n(t) + \mu(n+1)P_{n+1}(t)$$

$$n \geq 1 .$$

Suppose that the clone of cells becomes the detected tumor up to some predetermined size $N_{\max}$. In this case the probability density function of time growth, $\gamma(t)$, may be calculated from the condition

$$\gamma(t) = \beta \left(N_{\max} - 1 \right) P_{N_{\max}-1}(t) .$$

Deterministic Equation

In the case of the deterministic model we cannot write the density function of growth time. However, we can do the following. Assume that the tumor growth time, $t_{\max}$, is a constant. Then

$$r(t) = g \left(t - t_{\max} \right) .$$

If we assume that the cell population became the detected tumor with $N_{\max}$ cells, then the growth time may be calculated from the equation of tumor growth

$$z(t_{\max}, p) = N_{\max} ,$$

where $z(t, p)$ is a solution of the growth equation and p is a vector of parameters. For example, for the exponential growth model $N(t) = N_0 \exp(\lambda t)$, we have

$$t_{\max} = \frac{\ln(N_{\max}/N_0)}{\lambda} \, .$$

C.2 Cell-cycle Models

It is well known that during its lifetime a cell goes through a number of transformations. In mammalian cells, the division process is usually defined in relation to the two available biological markers of progress toward division. One of these markers is the occurrence of DNA synthesis. This can be recognized by cellular uptake of radiolabeled molecules which act as DNA precursors (^{3}H-thymidine). The phase during which DNA synthesis takes place is termed the S-phase. The proportion of labeled cells in tissue during a short interval (usually one hour) after the injection of ^{3}H-thymidine into an animal or patient is called the labeling index. It is denoted by LI and defined as $LI = qT_s/T_c$, where T_s and T_c are the duration of DNA synthesis and the cell-cycle time respectively. The term q, where $q > 0$, depends on the nature of the cell population.

The second marker occurs when the cell actually divides – one cell becoming two in a highly distinctive process, easily recognizable under the microscope and known as mitosis or the M-phase. In the simplest case (though there are others) each of the daughter cells now begins to prepare for cell division. These newly born cells don't immediately begin DNA synthesis, but go through a preparatory period (which is probably quite active in terms of gene switching and protein synthesis) which is referred to as the first of the gap phases and denoted by G_1. Similarly, a second gap (denoted by G_2) occurs between the S-phase and the onset of mitosis.

Cells do not necessarily spend all of their time traversing these four phases in a cyclical fashion. They can be switched out of the cycle into a so-called resting phase denoted by G_0 (see *Figure C.4*).

Multicompartmental models based on cell kinetics are used to study the problems associated with tumor growth and to construct an algorithm to find the best policy for chemotherapy. This work is normally based on the following factors:

- There are a wide range of drugs which act specifically in one or more cycle phases.

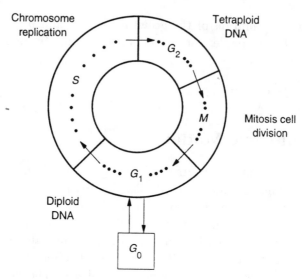

Cell differentiation or loss

Figure C.4. The cell-cycle concept: the process of one cell becoming two entails stepwise progression through a sequence of distinct phases after which the cell either enters a "resting" phase or re-enters the division process.

- Cancer and healthy cells spend different proportions of the cycle time in each phase.
- Cancer cells take longer to recover from drug effects than normal ones (Mendelsohn, 1977).

Sundareshan and Fundakowski (1986) show that most of the models of cell proliferation kinetics proposed to date are equivalent and consider a continuous time, multicompartmental model as the most adequate one for considering the effects of drugs on the cell cycle. By assigning more than one compartment to each phase we may take into account that, on average, each cell may spend different periods of time in the various phases. There are formulae for determining the number of compartments for each phase by analyzing DNA histogram data.

The typical model of this type can be written in the following form (Lobo Pereira *et al.*, 1990). Let N, the number of compartments, be

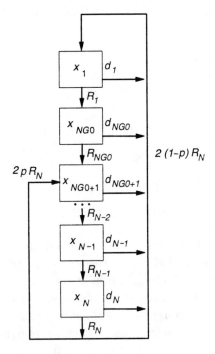

Figure C.5. Cell-cycle model.

given by $N = N_{G0} + N_{G1} + N_S + N_{G2} + N_M$, and let $N_{(.)}$ be the number of compartments in the phase (.). Then,

$$\frac{dx_i}{dt} = -(R_i + d_i)x_i(t) + R_{i-1}x_{i-1}(t) \qquad (C.4)$$

when $1 < i \le N_{G0}$ or $N_{G0} + 1 < i < N$, and

$$\begin{aligned}
\frac{dx_1(t)}{dt} &= -(R_1 + d_1)x_1(t) + 2(1-p)R_N x_N(t) \\
\frac{dx_{NG0+1}(t)}{dt} &= -(R_{NG0+1} + d_{NG0+1})x_{NG0+1} \\
&\quad + 2pR_N x_N(t) \ .
\end{aligned} \qquad (C.5)$$

Here, d_i is the death rate in the ith compartment, R_i is the transition rate from the ith to the $(i+1)$th compartment, and p is the fraction of cells that goes to phase G_1 after mitosis. Let Φ denote the following matrix (see *Figure C.5*):

$$\Phi = \begin{bmatrix} -(R_1 + d_1) & 0 & 0 & \cdots & & 2R_N(1-p) \\ R_1 & -(R_2 + d_2) & 0 & \cdots & & 2R_N p \\ 0 & R_2 & -(R_3 + d_3) & \cdots & & 0 \\ & & & \cdots & & \\ 0 & 0 & 0 & \cdots & -(R_{N-1} + d_N) & 0 \\ 0 & 0 & 0 & \cdots & R_{N-1} & -(R_N + d_N) \end{bmatrix}$$

Now, we can write

$$\frac{dX(t)}{dt} = \Phi X(t), \quad X(0) = X^0, \quad t \geq 0 \ .$$

The data used to estimate the parameters R_i, d_i, $i = 1, \ldots, N$, and p of matrix Φ consist of the relative number of cells in each compartment of a population which was initially synchronized at a certain point in time. The fact that current measurement techniques (namely microfluorometry with different markers) only produce the relative (to the total population) number of cells in each compartment implies that a fractional cell distribution model is required.

It is potentially useful to note again here the bilinear system structure of the cell-cycle kinetics given in equations (C.4) and (C.5) where R_i may be manipulated by appropriate treatment or therapy.

C.3 Tumor Growth Under Stochastic Environmental Conditions

The effect of environmental fluctuations on the macroscopic properties of non-linear systems has been investigated primarily in population dynamics (Lewontin and Chen, 1969; Levins, 1969a,b; May, 1973; Smith and Tuckwell, 1974). These studies have established that such fluctuations have a profound impact on the growth and extinction of populations and thus are important for the stability of systems. In biological processes there are two main sources of variability (Rosenkranz, 1985):

- Demographic stochasticity, which is due to randomness in the survival ability or fertility of individuals within a population.
- Environmental stochasticity, which results from random fluctuations in the environment affecting the population as a whole.

In this section a tumor growth model originally proposed by Garay and Lefever (1987) is considered. A stochastic variant of this model was studied by Lefever and Horsthemke (1979) and Rosenkranz (1985). The models describe three principal phenomena:

- The transformation of normal cells into neoplastic ones (at a constant rate A).
- The replication of transformed cells (at some rate λ).
- The immunological interaction of the host organism with transformed cells.

Within this framework, tumor growth is given by the equation

$$\frac{dz}{d\tau} = (N - z)\left(A + \frac{\lambda}{N}z\right) - \frac{k_1 E z}{1 + (k_1/k_2)z} \, , \tag{C.6}$$

where z is the population of target cells (tumor cells), E is the total number of effector cells, and $E = E_0 + E_1$, where E_0 is the population of the tree effector cells and E_1 is the population of effector cells which recognized and bound a target cell. The term N is the maximum number of target cells which can be packed in a given volume element. The terms A, λ, k_1, and k_2 are rate constants, and τ is normalized time.

Equation (C.6) describes the local transition mechanism (involving the local interactions of target and effector cells) between tissue states of different natures, and not the process of growth of the tumor as a whole. Thus, the factor $(N - z)$ expresses simply the existence of an upper limit N to the population z in the volume element under consideration. This N must not be confused with the saturation level in the equations of the logistic or Gompertz type which describe total populations. Some estimates of model parameters can be found in *Table C.1*.

The average rate of the transformation of normal cells into neoplastic ones (A) is very low compared with the rate of neoplastic replication. A minimal value can be estimated from the incidence of cancer in populations: this yields 10^{-17}–10^{-18} transformed cells/normal cell day. The growth constant, λ, can be determined from the experimental data on tumor growth. The range is given by $0.2 < \lambda < 1.5$ day^{-1}. The values of k_1 and k_2 are very variable, depending on the nature of the effector cells, on the tumor, on the level of sensitization, etc. From *Table C.1*, we can see that the scale of increasing cytotoxic activity for different, active effector cells is: activated macrophages < immune T lymphocytes and natural killer cells < allosensitized T lymphocytes.

Table C.1. Representative values of cytotoxic and rejection parameters for each kind of effector cell.

Effector cell	$k_1 N$ (day^{-1})	k_2 (day^{-1})	θ^a	$E_c{}^b$
Allosensitized T lymphocytes	18	18	1	0.01–0.1
Immune T lymphocytes	0.43	0.85	1–2	0.5–3.0
Syngenic natural killer cells	>1–6	0.6–3	<0.1–3	<0.3–1.3
Syngenic activated macrophages	0.1–0.4	0.2–0.7	0.5–5.0	0.6–10

^{a} The term θ is given by $\lambda k_2/k_1(\lambda - A)N$.
^{b} E_c is the minimum effector/target ratio for obtaining tumor rejection.
Source: Garay and Lefever, 1978.

In dimensionless form, equation (C.6) can be written as

$$\frac{dx}{dt} = \alpha + (1 - \theta x)x - \frac{\beta x}{1 + x}$$

$$\alpha = \frac{k_1 A N}{k_2(\lambda - A)}, \quad \beta = \frac{k_1 E_1}{\lambda - A}, \quad \theta = \frac{\lambda k_2}{k_1(\lambda - A)N}$$

$$t = (\lambda - A)\tau, \quad x = \frac{k_1}{k_2}z \ . \tag{C.7}$$

The corresponding ranges of the variation in the parameters α, β, and θ are $10^{-19} < \alpha < 10^{-16}$, $10^{-2} < \beta < 10$, $10^{-1} < \theta < 5$. The steady states of equation (C.6) are the solutions of the polynomial

$$-\theta x^3 + (1 + \theta)x^2 + (1 + \alpha - \beta)x + \alpha = 0 \ .$$

For the values of β within the range $10^{-2} < \beta < 10$, there are three physically meaningful, steady-state values for x if $\theta < 1$ and $\alpha < \alpha_c \equiv (1 - \theta)^3/27\theta^2$. There is a triple root corresponding to the values $\alpha = \alpha_c$, $\beta = \beta_c = (1 + 2\theta)^3/27\theta^2$, $x = x_c = (1 - \theta)/3\theta$. Assuming $x \ll 1$ and $\theta \ll 1$, bistability occurs in the interval $\beta_1 \approx (1 + \sqrt{\alpha})^2 < \beta < (1 - \theta)^2/4\theta \approx \beta_2$.

It is especially interesting to study the influence of external fluctuations on this dynamic model: such situations occur when β and/or λ fluctuate. Consider the case when the fluctuations around the mean value of β vary much more rapidly than the macroscopic evolution of x. The correlation time of the fluctuations leads to zero, i.e., we make the idealization of white noise with $E\beta_t = \beta$, $E(\beta_t - \beta)(\beta_t - \beta) = \sigma^2\delta(t - t')$. Now we can write the SDE associated with equation (C.6):

$$dx_t = \left(\alpha + (1 - \theta x)x - \frac{\beta x}{1 + x} \right) dt + \frac{\sigma x}{1 + x} dw_t$$

$$= f(x)dt + G(x)dw_t .\qquad (C.8)$$

To analyze the stationary solutions of equation (C.8) we write the Fokker–Plank equation

$$\frac{\partial p(x)}{\partial t} = -\frac{\partial}{\partial x} \left[\left(\alpha + (1 - \theta x)x - \beta\frac{x}{1 + x} \right) p(x) \right]$$
$$+ \frac{\sigma^2}{2}\frac{\partial^2}{\partial x^2} \left[\left(\frac{x}{1 + x} \right)^2 p(x) \right]$$

whose stationary solution is (for $\alpha = 0$)

$$p_s(x) = N \exp\frac{2}{\sigma^2} \left((2 - \theta - \beta)x - \frac{1}{2}(1 - 2\theta)x^2 \right.$$
$$\left. - \frac{\theta}{3}x^3 + (1 - \beta - \sigma^2) \log x + \sigma^2 \log(1 + x) \right) ,$$

which exists provided the following normalizability condition is satisfied:

$$\int_a^b p_s(x)\,dx < \infty .$$

Obviously x_t has to be positive so that $a = 0$ and $b = \infty$.

It seems very possible that in such complex in vivo systems the occurrence of instabilities and non-equilibrium phase transition phenomena may not be fully or reliably understood in strictly deterministic terms. The nature of fluctuations determines not only quantitatively the location of coexistence points in bistable systems but, furthermore, in the case of multiplicative noise, it induces transition phenomena which are absent from the deterministic equations.

The effect of noise can act in favor of cancer rejection as it can result in fluctuations in cytotoxic activity as well as in the replication rate. One might thus conjecture that noise effects could enhance the efficiency of immune defence mechanisms. Quite remarkably, this effect is obtained without varying the average value of the fluctuating parameters. One might speculate further that the therapeutic implications derived from this observation are that the administration of cytotoxic drugs or radiation amounts to multiplicative control on the growth parameters of a tumor.

References for Part III

Armitage, P., and Doll, R., 1954, *British Journal of Cancer* **8**:1–12.

Asachenkov, A.L., 1990, *Models and Methods for Processing Clinical Data from Oncological Patients* (in Russian), Department of Numerical Mathematics, Academy of Sciences of the USSR, Moscow, Russia.

Asachenkov, A.L., and Grigorenko, N., 1992, *Application of Control Theory in Cancer Research*, WP-92-56, International Institute for Applied Systems Analysis, Laxenburg, Austria.

Asachenkov, A.L., and Sobolev, B.G., 1993, *Parameter Estimation for the Survival Model*, WP-93-1, International Institute for Applied Systems Analysis, Laxenburg, Austria.

Asachenkov, A.L., Sobolev, B.G., Martianov, I.N., and Kurkin, A.N., 1987, *Index of Organ Defeat by Tumor* (in Russian), Preprint No. 115, Institute of Numerical Mathematics, Moscow, Russia.

Asachenkov, A.L., Sobolev, B.G., Martianov, I.N., and Kurkin, A.N., 1988, *Using the Index of Organ Defeat for the Analysis of the Tumor Growth Process* (in Russian), Preprint No. 176, Institute of Numerical Mathematics, Moscow, Russia.

Asachenkov, A.L., Sobolev, B.G., and Smolianinov, E.S., 1989, *Mathematical Modeling and Analysis of Data from Immunological Tests for Oncological Patients*, WP-89-32, International Institute for Applied Systems Analysis, Laxenburg, Austria.

Asherson, G.L., Colizzi, V., and Zembala, M., 1986, An Overview of T-Suppressor Cell Circuits, *Annual Review of Immunology* **4**:37–68.

Barrett, J.T., 1988, *Textbook of Immunology: An Introduction to Immunochemistry and Immunobiology*, 5th edn, The C.V. Mosby Co., St. Louis, MN, USA.

Benjamini, E., and Leskowitz, S., 1988, *Immunology: A Short Course*, A.R. Liss, Inc., New York, NY, USA.

Berzofsky, J.A., Brett, S.J., Streicher, H.Z., and Takahashi, H., 1988, Antigen Processing for Presentation to T Lymphocytes: Function, Mechanisms, and Implications for the T-Cell Repertoire, *Immunological Reviews* **106**:5–31.

De Boer, R. J., and Hogeweg, P., 1986, Interactions Between Macrophage and T-Lymphocytes: Tumor Sneaking – through Intrinsic to Helper T Cell Dynamics, *Journal of Theoretical Biology* **120**:331–351.

Bylov, B.F., Vinograd, D.M., and Grobman, V.V., 1966, *Theory of Lyapunov's Function* (in Russian), Nauka, Moscow, Russia.

Cameron R., McIntosh, J., and Rosenberg, S., 1988, Synergistic Antitumor Effects of Combination Immunotherapy with Recombinant Interleukin-2 and a Recombinant Hybrid A-Interferon in the Treatment of Established Murine Hepatic Metastases, *Cancer Research* **48**:5810–5817.

Carson, E.R. *et al.*, 1983, *The Mathematical Modelling of Metabolic and Endocrine Systems: Model Formulation, Identification and Validation*, John Wiley & Sons, New York, NY, USA.

Cox, D.R., and Oakes, D., 1984, *Analysis of Survival Data*, Chapman and Hall, London, UK.

Dilman, V.M., 1983, *Four Models of Medicine*, Meditsina, St. Petersburg, Russia.

Dohlsen, M. *et al.*, 1985, Proliferation of Human CD4+45R+ and CD4+45R– T Helper Cells is Promoted by both IL-2 and IL-4 while Interferon–Gamma Production is Restricted to IL-2 Activated CD4+45R– T Cells, *Immunological Letters* **20**:29–34.

Durum, S.K., 1985, Interleukin-1: An Immunological Perspective, *Annual Review of Immunology* **3**:263–287.

Erb, P., Ramila, G., Stern, A., and Sklenar, I., 1984, Role of Macrophages in T Cell Activation, pp. 17–22 in A.Ü. Muftüoğlu and N. Barlas, eds., *Recent Advances in Immunology*, Plenum Press, New York, NY, USA.

Van Furth, R. *et al.*, 1987, The Current View on the Origin of Pulmonary Macrophage, *Pathology Research and Practical* **175**:38–49.

Garay, R.P., and Lefever, R., 1987, A Kinetic Approach to the Immunology of Cancer: Stationary State Properties of Effector–Target Cell Reactions, *Journal of Theoretical Biology* **73**:417–438.

O'Garra, A., Umlard, S., De France, T., and Christiansen, J., 1988, B-Cell Factors are Pleiotropic, *Immunology Today* **9**(2):45–57.

Glass, L., and Mackey, M.C., 1979, Pathological Conditions Resulting from Instabilities in Physiological Control Systems, *Annals of the NY Academy of Sciences* **316**:214–235.

Gyllenberg, M., and Webb, G.F., 1989, *Quiescence as an Explanation of Gompertzian Tumor Growth*, Preprint: Research Report A264, Helsinki University of Technology, Institute of Mathematics, Helsinki, Finland.

Grossman, Z., and Berke, G., 1980, Tumor Escape from Immune Elimination, *Journal of Theoretical Biology* **83**:267–296.

Gronberg, A., 1989, Interferon is Able to Reduce Tumor Cell Susceptibility to Human Lymphokine-Activated Killer (LAK) Cells, *Cellular Immunology* **118**:10–21.

Hanson, F., and Tier, C., 1982, A Stochastic Model of Tumor Growth, *Mathematics of Bioscience* **61**:73–100.

Herberman, R.B., 1987, Activation of Natural Killer (NK) Cells and Mechanism of their Cytotoxic Effects, pp. 275–283 in S. Gupta *et al.*, eds., *Immune Regulation*, Plenum Press, New York, NY, USA.

Hill, R.P., 1987, Cancer as a Cellular Disease, in I.F. Tannock and R.P. Hill, eds., *The Basic Science of Oncology*, Pergamon Press, Oxford, UK.

Hooton, J.W.L., Gibbs, C., and Paetkau, V., 1985, Interaction of Interleukin 2 with Cells: Quantitative Analysis of Effects, *Journal of Immunology* **135**(4):2464–2473.

Iverson, S., and Arley, N., 1950, *Acta Pathologica Microbiologica Scandanavia* **27**:773–803.

Karlin, S., and Taylor, H.M., 1975, *A First Course in Stochastic Processes*, Second edn, Academic Press, New York, San Francisco, London.

Kevrekidis, I.G., Zecha, A.D., and Perelson, A.S., 1988, Modeling Dynamical Aspects of the Immune Response: 1. T Cell Proliferation and the Effect of IL-2, pp. 167–197 in A.S. Perelson, ed., *Theoretical Immunology*, Vol. 2, Addison-Wesley Publishing Co., Inc., Reading, MA, USA.

Klawansky, S., and Fox, M.S., 1984, *Journal of Theoretical Biology* **111**: 531–587.

Klein, E., and Vanky, F., 1981, Natural and Activated Cytotoxic Lymphocytes which Act on Autologous and Allogeneic Tumor Cells, *Cancer Immunology and Immunotherapy* **11**:183–188.

Laird, A.K., 1964, Dynamics of Tumor Growth, *British Journal of Cancer* **18**:490–502.

Lefever, R., and Horsthemke, W., 1979, Bistability in Fluctuating Environments: Implications in Tumor Immunology, *Bulletin of Mathematical Biology* **41**:469–490.

Lefever, R., and Garay, R., 1978, A Mathematical Model of the Immune Surveillance Against Cancer, pp. 481–518 in G.I. Bell *et al.*, eds., *Theoretical Immunology*, Marcel Dekker, New York, NY, USA.

Lewontin, R.C., and Chen, D., 1969, On Population Growth in a Randomly Varying Environment, *Proceedings of the National Academy of Sciences of the USA* **62**:1056–1060.

Levins, R., 1969a, Some Demographic and Genetic Consequences of Environmental Heterogeneity for Biological Control, *Bulletin of the Entomological Society of America* **15**:237–240.

Levins, R., 1969b, The Effect of Random Variations of Different Types on Population Growth, *Proceedings of the National Academy of Sciences of the USA* **62**:1061–1065.

Lobo Pereira, F.M.F., Pinho, M.R.M., Pedreira, C.E., and Fernandes, M.H.R., 1990, Design of Multidrug Cancer Chemotherapy via Optimal Control, pp. 39–44 in *Preprints of 11th IFAC World Congress*, Tallinn, Estonia, USSR, August 13–17, Vol. 4.

Lotze, M., and Rosenberg, S., 1988, The Immunological Treatment of Cancer, *A Cancer Journal for Clinicians* **38**(2):68–94.

Mackey, M.C., and Glass, L., 1977, Oscillations and Chaos in Physiological Control Systems, *Science* **197**:287–289.

Marchuk, G.I., Asachenkov, A.L., Sobolev, B.G., and Smolianinov, E.S., 1989, On the Problem of Analyzing Clinical Data from Oncological Patients, *Soviet Journal of Numerical Analysis and Mathematical Modelling* **4**(5):356–381.

Manton, K.G., and Stallard, E., 1984, *Recent Trends in Mortality Analysis*, Academic Press, Orlando, FL, USA.

May, R.M., 1973, *Stability and Complexity in Model Ecosystems*, Princeton University Press, Princeton, NJ, USA.

Mendelsohn, M., 1977, Principles, Relative Merits and Limitations of Current Cytokinetic Techniques, pp. 101–112 in B. Drewinko and R.M. Humphrey, eds., *Growth Kinetics and Biochemical Regulation of Normal and Malignant Cells*, Williams and Wilkins, Baltimore, MD, USA.

Merrill, S.J., 1983, A Model of the Role of Natural Killer Cells in Immune Surveillance – II, *Journal of Mathematical Biology* **17**:153–162.

Meager, A., Leung, H., and Woolley, J., 1989, Assays for Tumor Necrosis Factor and Related Cytokines, *Journal of Immunological Methods* **116**:1–17.

Milanese, C., Siliciano, R.F., Richardson, N.E., Chang, H.-C., Alcover, A., and Reinherz, E.L., 1987, Human T Lymphocyte Activation, pp. 59–67 in S. Gupta *et al.*, eds., *Mechanisms of Lymphocyte Activation and Immune Regulation*, Plenum Press, New York, NY, USA.

Mizel, S.B., 1982, Interleukin 1 and T Cell Activation, *Immunological Review* **63**:51–72.

Mohler, R.R., and Lee, K.S., 1990, On Tumor Modeling and Control, pp. 71–82 in R. Mohler and A. Asachenkov, eds., *Selected Topics on Mathematical Models in Immunology and Medicine*, CP-90-7, International Institute for Applied Systems Analysis, Laxenburg, Austria.

Murray, J.D., 1989, *Mathematical Biology*, Springer-Verlag, Berlin, Heidelberg, New York.

Muller, H.J., 1951, Radiation Damage to Genetic Material, pp. 93–165 in G.A. Baitsell, ed., *Science in Progress*, Vol. 7, Yale University Press, New Haven, CT, USA.

Nordling, C.O., 1954, New Theory on Cancer-inducing Mechanisms, *British Journal of Cancer* **7**:68–72.

Norton, L., and Simon, R., 1979, New Thoughts on the Relationship of Tumor Growth Characteristics to Sensitivity to Treatment, pp. 53–59 in V.T. De Vita, Jr., and H. Busch, eds., *Methods in Cancer Research*, **XVII**, Academic Press, New York, San Francisco, London.

Otter, W.D., 1986, The Effect of Activated Macrophage on Tumor Growth in Vitro and in Vivo, *Lymphokines* **3**:389–422.

Peto, R., Roe, F.C.J., Lee, P.N., Levy, L., and Clark, J., 1975, *British Journal of Cancer* **32**:411–426.

Rescigno, A., and DeLisi, C., 1977, Immune Surveillance and Neoplasia – II: A Two Stage Mathematical Model, *Bulletin of Mathematical Biology* **39**:487–497.

Rosenberg, S.A., Schwarz, S.L., and Spiess, P.J., 1988, Combination Immunotherapy for Cancer: Synergistic Antitumor Interactions of Interleukin-2, Alfa Interferon, and Tumor-Infiltrating Lymphocytes, *Journal of the National Cancer Institute* **80**(17):1393–1397.

Rosenkranz, G., 1985, Growth Models with Stochastic Differential Equations: An Example from Tumor Immunology, *Mathematical Biosciences* **75**: 175–186.

Scollay, R., and Shortman, K., 1985, Cell Traffic in Adult Thymus: Cell Entry and Exit, Cell Birth and Death, pp. 3–30 in J.D. Watson *et al.*, eds., *Recognition and Regulation in Cell-Mediated Immunity*, Marcel Dekker, Inc., New York and Basel.

Smith, C.E., and Tuckwell, H.C., 1974, Some Stochastic Growth Processes, pp. 211–225 in S. Levins, ed., *Lecture Notes in Biomathematics*, Vol. 2, Springer-Verlag, Berlin, Heidelberg, New York.

Steel, G.G., 1977, *Growth Kinetics of Tumors*, Clarendon Press, Oxford, UK.

Stocks, P., 1953, Study of Age Curve for Cancer of the Stomach in Connection with Theory of Cancer Producing Mechanisms, *British Journal of Cancer* **7**:407–417.

Sundareshan, M.K., and Fundakowski, 1986, Stability and Control of a Class of Compartmental System with Application to Cell Proliferation and Cancer Therapy, *IEE Transactions on Auto Control* **AC-31**(11):1022–1032.

Swan, G.W., 1984, *Applications of Optimal Control Theory in Biomedicine*, Marcel Dekker, Inc., New York and Basel.

Tannock, I.F., 1987, Tumor Growth and Cell Kinetics, pp. 160–170 in I.F. Tannock and R.P. Hill, eds., *The Basic Science of Oncology*, Pergamon Press, Oxford, UK.

Trinchieri, G. *et al.*, 1987, Regulation of Activation and Proliferation of Human Natural Killer Cells, pp. 285–298 in S. Gupta *et al.*, eds., *Mechanisms of Lymphocyte Activation and Immune Regulation*, Plenum Press, New York, NY, USA.

Ventcel, A.D., and Freidlin, M.I., 1979, *Fluctuations in Dynamical Systems Under Action of Small Random Perturbations* (in Russian), Nauka, Moscow, Russia.

Vilcek, J. *et al.*, 1985, Interleukin-2 as the Inducing Signal for Interferon-Gamma in Peripheral Blood Leukocytes Stimulated with Mitogen or Antigen, pp. 385–396, in H. Kichner and H. Schellekens, eds., *The Biology of the Interferon System, 1984*, Proceedings of the 1984 TNO–ISIR Meeting on the Biology of the Interferon System, held in Heidelberg, Germany, 21–25 October 1984, Elsevier, Amsterdam, The Netherlands.

Warga, J., 1972, *Optimal Control of Differential and Functional Equations*, Academic Press, New York, San Francisco, London.

Watson, G.S., 1977, *Proceedings of the National Academy of Sciences* **74**: 1341–1342.

Wheldon, T.E., 1988, *Mathematical Models in Cancer Research*, Adam Hilger, Bristol and Philadelphia.

Whittemore, A.S., and Keller, J.B., 1978, *Soc. Ind. Appl. Math. Rev.* **20**:1–30.

Woodbury, M.A., and Manton, K.G., 1977, A Random-Walk Model of Human Mortality and Aging, *Theoretical Population Biology* **11**:37–48.

Wright, S., 1926, Book Review, *Journal of the American Statistical Association* **21**:493–497.

Yashin, A.I., 1984, *Dynamics in Survival Analysis*, International Institute for Applied Systems Analysis, Laxenburg, Austria.

Yashin, A.I., 1985, Dynamics in Survival Analysis: Conditional Gaussian Property Versus the Cameron–Martin Formula, pp. 467–485 in N.V. Krylov, R.S. Lipster, and A.A. Novicov, eds., *Sleklov Seminar*, Springer-Verlag, Berlin, Heidelberg, New York.

Yashin, A.I., and Arjas, E., 1988, A Note on Random Intensities and Conditional Survival Function, *Journal of Applied Probability* **25**(3):630–635.

Yashin, A.I., Manton, K.G., and Stallard, E., 1986, Evaluating the Effects of Observed and Unobserved Diffusion Processes in Survival Analysis of Longitudinal Data, *Mathematical Modelling* **7**(9–12):1353–1363.

Zuev, S.M., 1988, *Statistical Estimating of Parameters of the Mathematical Models of Disease* (in Russian), Nauka, Moscow, Russia.

Systems & Control: Foundations & Applications

Series Editor
Christopher I. Byrnes
School of Engineering and Applied Science
Washington University
Campus P.O. 1040
One Brookings Drive
St. Louis, MO 63130-4899
U.S.A.

Systems & Control: Foundations & Applications publishes research monographs and advanced graduate texts dealing with areas of current research in all areas of systems and control theory and its applications to a wide variety of scientific disciplines.

We encourage the preparation of manuscripts in TeX, preferably in Plain or AMS TeX— LaTeX is also acceptable—for delivery as camera-ready hard copy which leads to rapid publication, or on a diskette that can interface with laser printers or typesetters.

Proposals should be sent directly to the editor or to: Birkhäuser Boston, 675 Massachusetts Avenue, Cambridge, MA 02139, U.S.A.

Estimation Techniques for Distributed Parameter Systems
H.T. Banks and K. Kunisch

Set-Valued Analysis
Jean-Pierre Aubin and Hélène Frankowska

Weak Convergence Methods and Singularly Perturbed
Stochastic Control and Filtering Problems
Harold J. Kushner

Methods of Algebraic Geometry in Control Theory: Part I
Scalar Linear Systems and Affine Algebraic Geometry
Peter Falb

H^{∞} - Optimal Control and Related Minimax Design Problems
Tamer Başar and Pierre Bernhard